Soil Remediation and Improvement through the Application of By-Product Amendments

Soil Remediation and Improvement through the Application of By-Product Amendments

Guest Editors

Elias Afif Khouri

Rubén Forján Castro

MDPI

Basel • Beijing • Wuhan • Barcelona • Belgrade • Novi Sad • Cluj • Manchester

Guest Editors

Elias Afif Khouri
Department of Organisms and Systems Biology
University of Oviedo
Mieres
Spain

Rubén Forján Castro
Department of Soil Science and Agricultural Chemistry and Environmental Geomorphology and Geological Heritage Research Group
University of Salamanca
Salamanca
Spain

Editorial Office

MDPI AG
Grosspeteranlage 5
4052 Basel, Switzerland

This is a reprint of the Special Issue, published open access by the journal *Environments* (ISSN 2076-3298), freely accessible at: https://www.mdpi.com/journal/environments/special_issues/2E2H5226D2.

For citation purposes, cite each article independently as indicated on the article page online and as indicated below:

Lastname, A.A.; Lastname, B.B. Article Title. *Journal Name* **Year**, *Volume Number*, Page Range.

ISBN 978-3-7258-5119-5 (Hbk)
ISBN 978-3-7258-5120-1 (PDF)
https://doi.org/10.3390/books978-3-7258-5120-1

Cover image courtesy of Rubén Forján Castro

Contents

About the Editors

Elias Afif Khouri

Elias Afif Khouri is a professor in the Department of Organisms and Systems Biology at the University of Oviedo in Asturias, Spain, and is affiliated with the Institute of Natural Resources and Territorial Planning (INDUROT). He teaches Soil Science, Hydraulics and Hydrology and Forest Hydrological Restoration in the Bachelor's Degree in Forestry and Sustainable Natural Resources Management at the Polytechnic School of Mieres. He is responsible for the Agroforestry Engineering Laboratory, where soil and plant samples are analyzed. His main research interests focus on soil science in relation to plant growth, with particular attention to soil fertility and remediation techniques for heavy metals in contaminated soils. His medium and long-term scientific goal is to generate new data through research projects and contracts that ensure the continuity of scientific production.

Rubén Forján Castro

Rubén Forján Castro is currently a researcher at SmartForest Research Group at Oviedo University. He is a lecturer and researcher at INDUROT and a promoter of the Pasek2.0 chair. He is currently involved in various European and national projects such as the LIFE Carbon2mine or Horizon Europe project C-SINK. Rubén researches the recovery of degraded soil and water. He has deep knowledge in compost, biochar, hydrochar, nanoparticles, phyoremediation, revalorization of waste, and soil carbon sequestration.

Preface

In the face of increasing soil degradation and the urgent need to find sustainable solutions accessible to our communities and territories, we have witnessed how organic and industrial waste—often seen as a problem—can become an opportunity to restore soil health and close natural cycles.

This Reprint is the result of a collective effort by researchers committed to the circular economy. It explores, from diverse perspectives and contexts, the potential of by-products as amendments for soil remediation and improvement. It is not only about scientific results, but also about concrete proposals that connect science with practice, reflecting the authors' commitment to more resilient agriculture, smarter waste management, and a more livable planet.

Beyond agronomic and environmental benefits, this collection highlights the importance of interdisciplinary approaches that integrate biotechnology, ecology, and sustainable waste management. The contributions presented here illustrate how science can offer practical and scalable solutions to complex problems.

As Guest Editors, it has been a privilege to accompany this process. We are deeply grateful to each author for their dedication, to the reviewers for their critical and constructive insights, and to the Environments editorial team for their constant support. We hope this Reprint not only informs, but also inspires: to continue researching, to collaborate across disciplines, and to apply knowledge for the benefit of our soils and our communities.

Elias Afif Khouri and Rubén Forján Castro
Guest Editors

environments

MDPI

Editorial

Soil Remediation and Improvement Through the Application of By-Product Amendments

Elias Afif [1] **and Rubén Forján** [2,3,*]

[1] Agroforestry Engineering Area, Department of Organisms and Systems Biology, Polytechnic School of Mieres, University of Oviedo, 33600 Mieres, Spain; elias@uniovi.es
[2] SMartForest Group, Department of Organisms and Systems Biology, University of Oviedo, 33600 Mieres, Spain
[3] Environmental Biogeochemistry & Raw Materials Group, Institute of Natural Resources and Territorial Planning (INDUROT), University of Oviedo, 33600 Mieres, Spain
* Correspondence: ruben.indurot@uniovi.es

Soils are fundamental to terrestrial ecosystems, human health, food production, and global environmental stability [1–4]. However, the combined threats of soil contamination and degradation—alongside growing pressures from industrial and agricultural waste production—require urgent and integrated responses [5–7]. This Special Issue of *Environments*, entitled "Soil Remediation and Improvement through the Application of By-Product Amendments", responds to this challenge by exploring how various waste-derived materials can be repurposed as soil amendments for both remediation and fertility enhancement [8–11].

The loss of productive soil due to contamination (e.g., from mining or industrial activities) and the simultaneous overproduction of organic and inorganic by-products represent two critical and interconnected environmental issues [12,13]. The aim of this Special Issue is to connect both problems by highlighting sustainable solutions based on the circular economy. The articles published here present innovative uses of by-products such as biochar, hydrochar, compost, digestate, ashes, and other organic materials for the rehabilitation of degraded soils (e.g., mine soils, brownfields) and the enhancement of soil quality in agricultural settings.

Among the contributions, Sinclair et al. (contribution 1) examine the composting of spent coffee grounds (SCGs) and its impact on sunflower germination and early development. Their study demonstrates that composted SCGs, when properly aged, can improve water retention and reduce phytotoxicity, supporting their reuse as soil amendments. Similarly, Huslina et al. (contribution 2) assess the phytoremediation of arsenic-contaminated mine waste using native Australian plant species supported by biosolid-derived biochar. Their results underscore the importance of selecting suitable amendment–plant combinations to optimize both biomass production and contaminant uptake.

Regarding organic waste valorization, Licitra et al. (contribution 3) provide an insightful case study combining the composting and vermicomposting of organic municipal waste in Sicily. By integrating *Eisenia fetida* earthworms, they achieved significant reductions in the carbon-to-nitrogen ratio and heavy metal content, demonstrating the potential of this approach for nutrient recycling and soil health restoration. De Carolis et al. (contribution 4) further show that digestate derived from livestock waste enhances lettuce biomass and photosynthetic performance, even under low copper stress, pointing toward the efficacy of anaerobic digestion by-products in sustainable horticulture.

The Issue also includes studies highlighting the effects of by-product amendments on soil microbial communities and nutrient cycling. Deinert and Schmalenberger (contribution 5)

 https://doi.org/10.3390/environments12050170

demonstrate that ash-derived fertilizers stimulate microbial phosphorus mobilization in the rhizosphere of *Lolium perenne*, offering an alternative to conventional superphosphate fertilization. Works like this, due to their effectiveness, have been carried out by researchers such as Schiemenz et al. [14] and Doni et al. (contribution 6), who examine the combined use of zeolite and winery waste compost in vineyards, showing improvements in nutrient retention, carbon sequestration, and enzymatic soil activity—especially with a 10% zeolite blend.

In the area of sustainable crop production, Chanda et al. (contribution 7) present a compelling case for cyanobacteria biofertilizers in okra cultivation. Their experiments highlight comparable yield performance between cyanobacteria and synthetic fertilizers, while also promoting iron uptake and reducing chlorosis. Similarly, Carril et al. (contribution 8) evaluate the application of wood distillate (pyroligneous acid) to three leguminous crops, showing significant improvements in yield and mineral composition, particularly for lentils.

Finally, Sabia et al. (contribution 9) present a broader perspective by calculating the carbon footprint of livestock feed production using by-product-based concentrates. Their findings reveal that incorporating agricultural by-products can reduce carbon emissions by up to 37%, supporting the idea that soil amendment strategies can also contribute to climate mitigation goals when considered across the food system [15].

Collectively, the articles in this Special Issue represent valuable contributions to the literature on soil remediation, organic waste valorization, and sustainable land management [16–18]. They highlight that the use of by-products is not only environmentally sound but also agronomically effective, especially when paired with biotechnological and ecological strategies such as phytoremediation, vermicomposting, and microbial activation [19–21].

Beyond highlighting case studies within diverse agroecological and industrial contexts, the articles in this Special Issue collectively underscore a paradigm shift toward integrated and circular approaches to soil remediation and improvement through the application of by-product amendments. The novelty of individual contributions lies in their interdisciplinary exploration of how waste-derived materials—such as biochar, compost, digestate, ashes, pyroligneous acid, and vermicompost—can be strategically applied to degraded or nutrient-poor soils. These studies go beyond laboratory assessments to offer field-applicable evidence on improving soil structure, microbial function, plant productivity, and contaminant immobilization. Importantly, many contributions adopt a systems perspective that connects soil recovery to broader goals such as waste valorization, climate change mitigation, and sustainable agriculture. Future research should build on these findings by refining amendment combinations for site-specific challenges, monitoring long-term ecological outcomes, and scaling up successful practices within circular bioeconomy frameworks. Overall, this Special Issue posits leveraging by-product-based soil amendments as crucial to the transition toward more resilient and regenerative land use systems.

We thank all authors for their high-quality contributions and the reviewers for their rigorous assessments. It is our hope that this collection inspires further interdisciplinary research and practical applications at the intersection of waste management, soil science, and sustainable agriculture. The adoption of circular economy principles in soil restoration practices is not only timely—it is essential.

Conflicts of Interest: The authors declare no conflict of interest.

List of Contributions:

1. Sinclair, C.L.; Irga, P.J.; Duani, G.; Torpy, F.R. Spent Coffee Grounds (SCGs) as a Soil Amendment: The Effects of Composting Time on Early Sunflower Development. *Environments* **2024**, *11*, 272. https://doi.org/10.3390/environments11120272.
2. Huslina, F.; Khudur, L.S.; Besedin, J.A.; Nahar, K.; Shah, K.; Surapaneni, A.; Netherway, P.; Ball, A.S. The Phytoremediation of Arsenic-Contaminated Waste by *Poa labillardieri*, *Juncus pauciflorus*, and *Rytidosperma caespitosum*. *Environments* **2025**, *12*, 60. https://doi.org/10.3390/environments12020060.
3. Licitra, E.; Giustra, M.G.; Di Bella, G.; Messineo, A. Combination between Composting and Vermicomposting of OFMSW: A Sicilian Case Study. *Environments* **2024**, *11*, 183. https://doi.org/10.3390/environments11080183.
4. De Carolis, C.; Iori, V.; Narciso, A.; Gentile, D.; Casentini, B.; Pietrini, F.; Grenni, P.; Barra Caracciolo, A.; Iannelli, M.A. The Effects of Different Combinations of Cattle Organic Soil Amendments and Copper on Lettuce (cv. *Rufus*) Plant Growth. *Environments* **2024**, *11*, 134. https://doi.org/10.3390/environments11070134.
5. Deinert, L.; Schmalenberger, A. Reuse of Soils Fertilized with Ash as Recycling Derived Fertilizer Revealed Strong Stimulation of Microbial Communities Involved in P Mobilization in *Lolium perenne* Rhizospheres. *Environments* **2024**, *11*, 49. https://doi.org/10.3390/environments11030049.
6. Doni, S.; Masciandaro, G.; Macci, C.; Manzi, D.; Mattii, G.B.; Cataldo, E.; Gispert, M.; Vannucchi, F.; Peruzzi, E. Zeolite and Winery Waste as Innovative By-Product for Vineyard Soil Management. *Environments* **2024**, *11*, 29. https://doi.org/10.3390/environments11020029.
7. Chanda, S.; Dattamudi, S.; Jayachandran, K.; Scinto, L.J.; Bhat, M. The Application of Cyanobacteria as a Biofertilizer for Okra (*Abelmoschus esculentus*) Production with a Focus on Environmental and Ecological Sustainability. *Environments* **2024**, *11*, 45. https://doi.org/10.3390/environments11030045.
8. Carril, P.; Bianchi, E.; Cicchi, C.; Coppi, A.; Dainelli, M.; Gonnelli, C.; Loppi, S.; Pazzagli, L.; Colzi, I. Effects of Wood Distillate (Pyroligneous Acid) on the Yield Parameters and Mineral Composition of Three Leguminous Crops. *Environments* **2023**, *10*, 126. https://doi.org/10.3390/environments10070126.
9. Sabia, E.; Braghieri, A.; Vignozzi, L.; Paolino, R.; Cosentino, C.; Di Trana, A.; Pacelli, C. Carbon Footprint of By-Product Concentrate Feed: A Case Study. *Environments* **2025**, *12*, 42. https://doi.org/10.3390/environments12020042.

References

1. Forján, R.; Rodríguez-Vila, A.; Covelo, E.F. Using compost and technosol combined with biochar and Brassica juncea L. to decrease the bioavailable metal concentration in soil from a copper mine settling pond. *Environ. Sci. Pollut. Res.* **2018**, *25*, 1294–1305. [CrossRef] [PubMed]
2. Lü, Y.; An, B.; Pan, Q.; Liu, W.; Sun, J.; Wang, J.; Qi, Z.; Li, C.H.; Dou, S.; Han, X. Nutrient amendment promotes vegetation restoration and improves ecosystem carbon uptake capacity in a degraded grassland. *Agric. Ecosyst. Environ.* **2025**, *388*, 109666. [CrossRef]
3. Kopittke, P.M.; Menzies, N.W.; Wang, P.; McKenna, B.A.; Lombi, E. Soil and the intensification of agriculture for global food security. *Environ. Int.* **2019**, *132*, 105078. [CrossRef]
4. Rakshit, A.; Singh, S.K.; Abhilash, P.C.; Biswas, A. (Eds.) *Soil Science: Fundamentals to Recent Advances*; Springer Nature Singapore Pte Ltd.: Singapore, 2021.
5. Peña-Álvarez, V.; Asensio, V.; Baragaño, D.; Forján, R.; Peláez, A.I.; Gallego, J.L.R. Integrated landfarming strategy for remediation of HCH-contaminated soil: Synergistic effects of bioaugmentation, organic amendments, and nanoscale zero-valent iron. *J. Hazard. Mater.* **2025**, *489*, 137637. [CrossRef]
6. Chen, Y.; Wu, P.; Li, X.; Li, L. Remediation of antimony-arsenic co-contaminated soils in mining areas: Synergistic interaction of amendments and plants affects soil Sb and As bioavailability, nutrient and enzyme activity. *Environ. Technol. Innov.* **2025**, *38*, 104185. [CrossRef]
7. Hou, D.; Jia, X.; Wang, L.; McGrath, S.P.; Zhu, Y.G.; Hu, Q.; Zhao, F.J.; Bank, M.S.; O'Connor, D.; Nriagu, J. Global soil pollution by toxic metals threatens agriculture and human health. *Science* **2025**, *388*, 316–321. [CrossRef]

8. Hong, L.; Feng, S.; Li, P.; Wang, A. Ecological Restoration and Regeneration Strategies for the Gumi Mountain Mining Area in Wuhan Guided by Nature-Based Solution (NbS) Concepts. *Sustainability* **2025**, *17*, 1913. [CrossRef]
9. Sahlaoui, T.; Raklami, A.; Heinza, S.; Marschner, B.; Bargaz, A.; Oufdou, K. Nature-based remediation of mine tailings: Synergistic effects of narrow-leafed lupine and organo-mineral amendments on soil nutrient-acquiring enzymes and microbial activity. *J. Environ. Manag.* **2024**, *371*, 123035. [CrossRef]
10. Palansooriya, K.N.; Dissanayake, P.D.; Igalavithana, A.D.; Tang, R.; Cai, Y.; Chang, S.X. Converting food waste into soil amendments for improving soil sustainability and crop productivity: A review. *Sci. Total Environ.* **2023**, *881*, 163311. [CrossRef]
11. Skrzypczak, D.; Mironiuk, M.; Witek-Krowiak, A.; Mikula, K.; Pstrowska, K.; Łużny, R.; Mościcki, K.; Pawlak-Kruczek, H.; Siarkowska, A.; Moustakas, K.; et al. Innovative fertilizers and soil amendments based on hydrochar from brewery waste. *Clean Technol. Environ. Policy* **2024**, *26*, 1571–1586. [CrossRef]
12. Rahman, Z.; Singh, V.P. The relative impact of toxic heavy metals (THMs) (arsenic (As), cadmium (Cd), chromium (Cr)(VI), mercury (Hg), and lead (Pb)) on the total environment: An overview. *Environ. Monit. Assess.* **2019**, *191*, 419. [CrossRef] [PubMed]
13. Loska, K.; Wiechula, D.; Korus, I. Metal contamination of farming soils affected by industry. *Environ. Int.* **2004**, *30*, 159–165. [CrossRef] [PubMed]
14. Schiemenz, K.; Eichler-Löbermann, B. Biomass ashes and their phosphorus fertilizing effect on different crops. *Nutr. Cycl. Agroecosyst.* **2010**, *87*, 471–482. [CrossRef]
15. Zaman, M.; Kleineidam, K.; Bekken, L.; Berendt, J.; Bracken, C.; Butterbach-Bahl, K.; Cai, Z.; Chang, S.X.; Clough, T.; Dawar, K.; et al. Climate-Smart Agriculture Practices for Mitigating Greenhouse Gas Emissions. In *Measuring Emission of Agricultural Greenhouse Gases and Developing Mitigation Options using Nuclear and Related Techniques*; Zaman, M., Heng, L., Müller, C., Eds.; Springer: Cham, Switzerland, 2021. [CrossRef]
16. Xiong, X.; Liu, J.; Xiao, T.; Lin, K.; Huang, Y.; Deng, P.; Hu, H.; Wang, J. Remediation of uranium-contaminated water and soil by biochar-based materials: A review. *Biochar* **2025**, *7*, 41. [CrossRef]
17. Protogene, M.; Murindangabo, Y.T.; Frouz, J.; Brom, J. Characterization, fractionation and untapped potential of phosphate-amended sewage sludge biochar in soil-plant systems. *Chemosphere* **2024**, *367*, 143565. [CrossRef]
18. Salgado, L.; Aparicio, L.; Afif, E.; Fernández-López, E.; Gallego, J.R.; Forján, R. A second life for mining waste as an amendment for soil remediation. *J. Mater. Cycles Waste Manag.* **2024**, *26*, 2971–2979. [CrossRef]
19. Saikia, S.; Dansena, S.K.; Kalamdhad, A.S. Application of unsegregated municipal solid waste char produced at low temperature and vermicompost for remediation of coal mine tailing soil. *Environ. Geochem. Health* **2025**, *47*, 143. [CrossRef] [PubMed]
20. Hossain, M.Z.; Islam, M.A.; Kibria, K.Q.; Islam, M.A. Effect of organic amendments on the alleviation of cadmium in red amaranth (Amaranthus gangeticus) grown in Cd-contaminated saline soils. *Environ. Monit. Assess.* **2025**, *197*, 7. [CrossRef] [PubMed]
21. Zhang, T.; Wang, X.L.; Zhou, J.; Zhou, W.; Zhou, S.Q. Construction of Phosphate-Solubilizing Microbial Consortium and Its Effect on the Remediation of Saline-Alkali Soil. *Microb. Ecol.* **2025**, *88*, 11. [CrossRef] [PubMed]

environments

MDPI

Article

Effects of Wood Distillate (Pyroligneous Acid) on the Yield Parameters and Mineral Composition of Three Leguminous Crops

Pablo Carril [1], Elisabetta Bianchi [1,2,*], Costanza Cicchi [3], Andrea Coppi [2], Marco Dainelli [2], Cristina Gonnelli [2], Stefano Loppi [1,4], Luigia Pazzagli [3] and Ilaria Colzi [2]

[1] Department of Life Sciences, University of Siena, Via Mattioli 3, 53100 Siena, Italy; pablo.carril@unisi.it (P.C.); stefano.loppi@unisi.it (S.L.)

[2] Department of Biology, University of Florence, Via Micheli 1, 50121 Florence, Italy; andrea.coppi@unifi.it (A.C.); marco.dainelli@unifi.it (M.D.); cristina.gonnelli@unifi.it (C.G.); ilaria.colzi@unifi.it (I.C.)

[3] Department of Experimental and Clinical Biomedical Sciences "Mario Serio", University of Florence, Viale Morgagni 50, 50134 Florence, Italy; costanza.cicchi@stud.unifi.it (C.C.); luigia.pazzagli@unifi.it (L.P.)

[4] BAT Center—Interuniversity Center for Studies on Bioinspired Agro-Environmental Technology, University of Naples 'Federico II', 80055 Naples, Italy

* Correspondence: e.bianchi@unifi.it

Abstract: The excessive use of chemical fertilizers and pesticides in agriculture is increasing the demand for novel products to improve the quality of crops in a more sustainable way. Wood distillate (WD, pyroligneous acid) is a by-product obtained during the pyrolysis of plant biomass that can be successfully applied in agriculture due to its ability to enhance the growth, size, and weight of edible plant parts. However, there is little information concerning its plant yield-promoting effects on leguminous crops. The present work investigated the effects of WD on the yield, protein content and mineral composition of chickpea (*Cicer arietinum* L.), lentil (*Lens culinaris* L.) and bean (*Phaseolus vulgaris* L.) plants grown in field conditions. The application of WD showed remarkable yield-promoting effects mostly in lentil plants, which significantly increased plant and shoot biomass, the number and weight of both pods and seeds, as well as the total seed protein content. Furthermore, seeds from WD-treated plants differentially increased the concentration of elements with high nutritional value for human health, including Fe, Ca, Mg and K. These results suggest that the effects of WD among the legumes tested are species-specific and that WD could be an optimal candidate to grow high-yielding legumes with improved seed nutritional quality.

Keywords: wood distillate; chickpea; lentil; bean; seed quality; minerals; proteins; crop yield

1. Introduction

Breeding crops for 8 billion humans without impacting natural resources is one of the major challenges of current agriculture [1]. To keep pace with population growth, crop production has relied on the application of chemical fertilizers and pesticides. However, their widespread use has led to a rapid degradation of natural resources and the unsustainable use of soil and water [2]. Such critical environmental issues demand alternative measures to improve crop production without jeopardizing human health, farm animals and the environment.

In this context, the valorization of waste plant biomass is becoming a promising approach to find valuable and more eco-friendly products for agricultural application [3]. Recently, pyroligneus acid, also known as wood distillate (WD), was included in the list of products that can be used in organic farming in Italy [4]. WD is a by-product obtained from the distillation of gases produced during the pyrolysis of woody biomass [5]. It is composed of over 200 water-soluble compounds, including phenols, tannins, esters and acetic acids, which are co-related with both its plant growth- and defense-promoting effects [3]. The chemical composition of WD can vary depending on the type of wood (conifer or broadleaf) and plant species used as feedstock [6]. As a consequence, these factors also determine the specific effects of WD on plants [7].

The positive effects of WD have been demonstrated in both horticultural and cereal crops. The foliar application of WD has been shown to protect lettuce (*Lactuca sativa* L.) plants from ozone-induced damage and to positively affect not only biomass accumulation, but also its chlorophyll and sugar content [8–10]. In addition, the beneficial effects of WD have been linked to increased N and P content in basil (*Ocimum basilicum* L.) [11], to an enhanced number of fruits and increased elemental composition in tomato [12], to increased growth, fruit weight and sweetness in rockmelon [13], and to increased shoot growth and grain yield in rice (*Oryza sativa* L.) [14]. To the best of our knowledge, only one study has investigated the effects of WD in legumes [15], and this study showed that chickpea (*Cicer arietinum* L.) plants weekly sprayed with 0.25% (v/v) chestnut-derived WD increased both seed weight and diameter, as well as both seed antioxidant power and seed protein content [15].

Legumes are among the most important crops in agriculture, accounting for 27% of the world's primary food production [16]. They are an excellent source of proteins, carbohydrates, and minerals, such as potassium (K), calcium (Ca), magnesium (Mg), iron (Fe), and zinc (Zn), and are gaining preference over animal proteins for human consumption [17,18]. In addition, there is a growing recognition of the contribution of legumes to the critical targets under Sustainable Development Goal 2 set out by the FAO, particularly regarding food access, smallholder incomes, as well as sustainable and resilient agriculture [19,20]. However, the unavailability of an adequate number of high-quality seeds and the lack of sustainable growth-boosters are major constraints in legume cultivation, which is often not sufficient to obtain viable production, especially in developing countries [21,22].

Although WD may be a promising alternative to sustainably increase crop production, its effect across different legume plants has never been investigated. In light of the ecological and economic benefits of legumes, this study aims to investigate whether WD has positive effects on the yield parameters and the nutritional content of bean (*Phaseolus vulgaris* L.), chickpea (*C. arietinum* L.), and lentil (*Lens culinaris* L.), three of the most important leguminous crops for human consumption worldwide [23,24]. The present work expands scientific knowledge on the possible phytostimulatory effects of WD and its potential role in the sustainable improvement of legume production.

2. Materials and Methods

2.1. Plant Material, Growth Conditions and Yield Parameters

Bean (*Phaseolus vulgaris* L.), chickpea (*C. arietinum* L.), and lentil (*Lens culinaris* L.) seeds were kindly provided by Del Colle Srl. Chestnut (*Castanea sativa* Mill.). Wood distillate (WD, BioDea®, Arezzo, Italy) was chosen owing to its previously reported beneficial effects on chickpea plants [15]. The WD provided by the producer possessed the following characteristics: pH 3.5–4.5; acetic acid 2–2.3% (v/v); density 1.05 kg L^{-1}; and polyphenol content in the range of 22–25 g L^{-1}. The element concentration in pure WD was determined as described in Section 2.3. and it was as follows: Fe 3.2 ± 0.05 mg L^{-1}, Na 4.9 ± 0.4 mg L^{-1}, K 32.9 ± 0.6 mg L^{-1}, Ca 944.2 ± 5.3 mg L^{-1}, Zn 3.6 ± 0.1 mg L^{-1}, and Mg 16.0 ± 1.0 mg L^{-1}. Thus, approximately the following contents of nutritional elements were sprayed onto the leaves (mg): Fe 0.0006, Na < 0.001, K 0.006, Ca 0.018, Zn 0.0007 and Mg 0.0003; or were used for the fertigation (mg): Fe 0.001, Na > 0.001, K 0.009, Ca 0.02, Zn 0.001 and Mg 0.0004. Seeds of each legume species were sown in separate plots in a crop field of Del Colle Srl. located at Bientina (Pisa, Tuscany, Italy). Plants were grown in a soil characterized by 26% clay, 37% sand, 3.3% organic matter, 17 cmol Kg^{-1} of cation-exchange capacity and a pH of 7.9. No mineral fertilizers were added. Each plot (2.5 m $\times$ 4 m) consisted of six rows at 50 cm intervals, and twenty seeds per row were sown at 20 cm intervals (120 plants per each legume species; sowing density: 12 seeds/m^2). Approximately 100 mL of WD was applied for both foliar and fertigation treatments, previously diluting the WD in tap water to a 0.2% (v/v) concentration for foliar treatments and to 0.3% (v/v) for fertigation treatments, according to the producer's instructions. Plants from the first three rows had their leaves sprayed weekly and were fertigated every two weeks with WD. The other three rows were treated in the same way but using tap water only (control). Plants were grown for four months (April–July 2022), until they were dry. Both whole plants and shoots were individually weighed, and the number, mean weight and total weight of both pods and seeds were recorded.

2.2. Protein Quantification

For protein quantification, ten seeds from each plant were randomly selected and pooled. To extract and quantify proteins, dry seeds were crushed with a mortar and pestle and then ground with an Ultra Turrax (T 25 Stirrer ULTRA-TURRAX®, IKA-Werke GmbH & Co. KG, Staufen, Germany)) at 220 V and 170 W for 20 s, with 5 cycles. Then, 0.1 g of then fine powder obtained was weighed out and added to 1 mL of a glacial extracting solution 10% TCA/acetone, according to [25]. The proteins were left to precipitate overnight at −20 °C and then the samples were centrifuged at 13,200× *g* for 15 min at 4 °C. The pellet was washed twice with cold acetone and centrifuged again, then the supernatant was discarded, and the pellet was dried to remove residual acetone. The proteins extracted from the samples were resuspended in PBS (phosphate saline buffer, pH 7.4, 137 mM NaCl, 2.7 mM KCl, 8 mM, Na_2HPO_4, 2 mM KH_2PO_4) and finally they were quantified using the Bradford assay [26]. The results were reported as mg of protein in 100 mg of seeds.

2.3. Mineral Determination in the Seeds and in WD

For element quantification, ten seeds from each plant were randomly selected and pooled. Seeds were carefully washed with demineralized water, dried at 70 °C for two days, ground with a mortar and pestle, and then dried again. About 0.1 g of seed powder was mineralized with 10 mL of 69% HNO_3 in a microwave digestion system (Mars 6, CEM Matthews, NC, USA) with maximum temperature of 200 °C for 10 min [27]. In the case of WD, 20 mL of pure WD was mineralized with 1.5 mL of 69% HNO_3 and 1 mL 37% HCl, according to [28]. Samples were diluted at a 1/20 dilution ratio with 0.5% $LaCl_3$ in 1% HNO_3. The concentration of micro- and macro-elements (K, Ca, Mg, Fe, Zn, and Na) was measured by means of atomic absorption spectroscopy (PinAAcle 500, Perkin Elmer) and expressed on a dry weight basis in the case of seeds, and in mg/L in the case of WD. To verify the method's reliability and accuracy, certified reference materials (grade BCR, Fluka Analytical, Sigma-Aldrich) were used. Recoveries were within ±10% and the precision was >95%.

2.4. Statistical Analysis

The data approached a normal distribution (Shapiro–Wilk test, $p < 0.05$), and hence a Student's t-test for independent samples was used to check for statistically significant ($p < 0.05$) differences between plants treated with water (control) and those treated with WD.

3. Results

3.1. Yield Parameters

The application of WD positively affected all investigated yield parameters in *L. culinaris*. Both total and above-ground plant biomass in this species increased significantly (3.5 g in WD-treated vs. 2.15 g in control plants) (Figure 1a,b). In addition, WD-treated *L. culinaris* showed an increased number of both pods (37.4 in WD-treated vs. 23.5 in control) and seeds (50.6 in WD-treated vs. 29.6 in control) per plant (Figure 1c,f), as well as an increase in the total and mean weight of both pods (2.4 g in WD-treated vs. 0.9 g in control) and in the mean weight of pods (0.05 g in WD-treated vs. 0.03 g in control plants) (Figure 1d,e). Moreover, an increase in the total weight of seeds (2.11 g in WD-treated vs.

0.9 g in control) and the mean weight of seeds (0.04 g in WD-treated vs. 0.03 g in control) (Figure 1g,h) per plant was also recorded in *L. culinaris*. Regarding the other two species, only *P. vulgaris* showed a significant increase in the number of pods (3.6 in WD-treated vs. 2.5 in control).

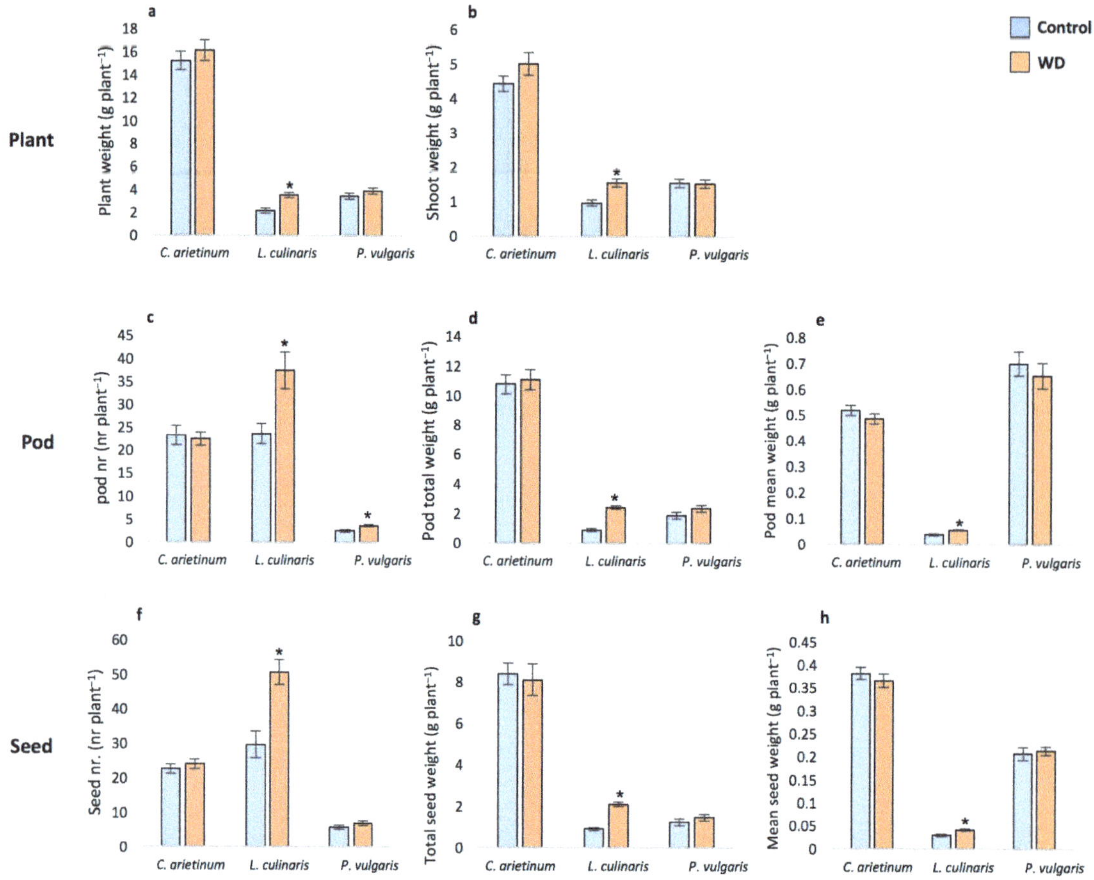

Figure 1. Yield parameters (mean ± standard error) of *C. arietinum*, *L. culinaris* and *P. vulgaris* plants treated with water (Control) or with 0.2% wood distillate (WD). (**a**) Plant fresh weight; (**b**) shoot fresh weight; (**c**) pod number; (**d**) total pod weight; (**e**) mean pod weight; (**f**) seed number; (**g**) total seed weight; (**h**) mean seed weight. * = Significant difference between control and WD-treated plants.

3.2. Total Proteins and Mineral Composition

Lentil seeds from WD-treated plants showed a significantly higher concentration of total proteins (17% in WD-treated vs. 15% in control), while no significant changes were found in *C. arietinum* and *P. vulgaris* (Figure 2a).

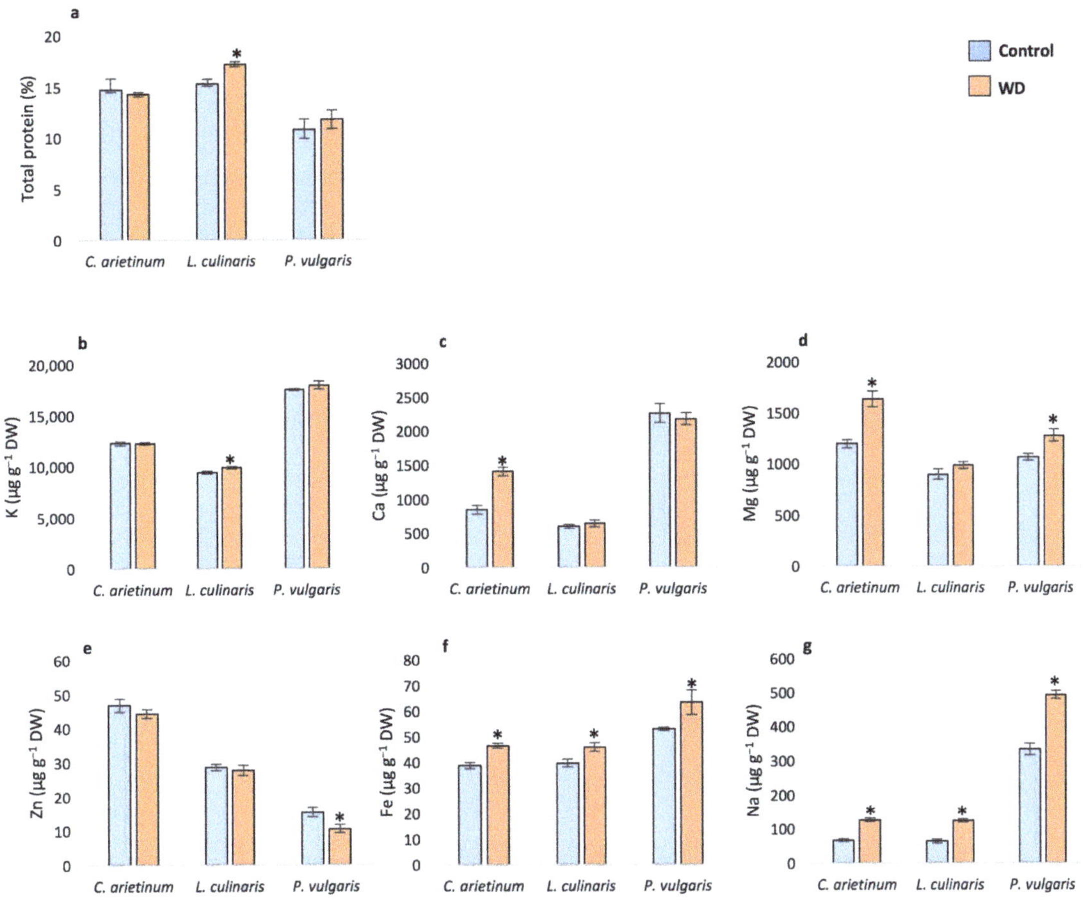

Figure 2. Total protein and mineral content (mean ± standard error) in *C. arietinum*, *L. culinaris* and *P. vulgaris* treated with water (Control) or with 0.2% wood distillate (WD). (**a**) Total protein concentration; (**b–g**) mineral concentration of (**b**) K, (**c**) Ca, (**d**) Mg, (**e**) Zn, (**f**) Fe, and (**g**) Na. * = significant difference between control and WD-treated plants.

Regarding the concentration of the investigated elements (Figure 2b–g), treatment with WD consistently increased the levels of Fe in the seeds of all species (46.5 µg g^{-1} DW in WD-treated vs. 38.7 µg g^{-1} DW in the control for *C. arietinum*, 45.8 µg g^{-1} DW in WD-treated vs. 39.7 µg g^{-1} in the control for *L. culinaris*; 63.2 µg g^{-1} DW in WD-treated vs. 52.8 µg g^{-1} DW in the control for *P. vulgaris*). The amount of Na also increased in the three species (126.8 µg g^{-1} DW in WD-treated vs. 67.4 µg g^{-1} DW in the control for *C. arietinum*; 124.4 µg g^{-1} DW in WD-treated vs. 64 µg g^{-1} in the control for *L. culinaris*; 493 µg g^{-1} DW in WD-treated vs. 333.2 µg g^{-1} DW in control for *P. vulgaris*). Furthermore, WD significantly increased the concentration in *C. arietinum* seeds (1634 µg g^{-1} in WD-treated vs. 1192 µg g^{-1} in the control for Mg and 1405.3 µg g^{-1} in WD-treated vs. 845.8 µg g^{-1} in the control for Ca). An increase in Mg levels was also observed in *P. vulgaris* (1276.2 µg g^{-1} in WD-treated vs. 1071.4 µg g^{-1} in control). Finally, the levels of K increased significantly only in *L. culinaris* (9866 µg g^{-1} in WD-treated vs. 9399.5 µg g^{-1} in control), while the concentration of Zn was reduced in WD-treated *P. vulgaris* (10.6 µg g^{-1} in WD-treated vs. 15.4 µg g^{-1} in control).

4. Discussion

Few studies have focused on the effects of WD application on legume crops grown in field conditions [29]. On the contrary, field experiments involving WD effects in the field have been performed with other plant species. For example, rice plants sprayed with 1/100 diluted WD both onto their leaves and onto the soil demonstrated an increased plant height and tiller production [30]. In addition, lettuce, cabbage, and cucumber plants that were leaf-sprayed with 500-fold diluted WD accumulated more vitamin C and demonstrated improved productivity [31]. More recently, Ref. [32] showed that rapeseed plants foliar sprayed with 1/400-fold diluted WD displayed a significantly increased seed yield, leaf area index, and number of pods. To the best of our knowledge, this is the second set of evidence reported for *C. arietinum* and the first set of evidence reported for both *L. culinaris* and *P. vulgaris* regarding the positive effects of WD on plant growth and yield. The yield-promoting effects of WD were mostly observed in *L. culinaris*—in addition to increasing total plant and shoot biomass, WD treatment almost doubled the number of seeds per plant and more than doubled their mean weight. As for *P. vulgaris*, the number of pods significantly increased in WD-treated plants, and a similar trend was observed for both seed number and weight, although this trend was not statistically significant. The observed WD-mediated increase in plant yield is consistent with previous reports focused on tobacco [33] as well as on both horticultural [9,10,34] and cereal [14] crops. In contrast, all yield parameters of *C. arietinum* were similar to the control, consistent with [15], although these authors observed a significant increase in the weight of WD-treated seeds. This might be due to the lower concentration of WD used in this study (0.2% vs. 0.25%), as well as its lower application frequency (every two weeks vs. every week). The different yield effects found between these three legumes indicate that the effects of WD are likely species-specific and that the concentrations tested are optimal to improve *L. culinaris* productivity. In this regard, higher WD concentrations may be tested for *P. vulgaris* and *C. arietinum* in future studies. In addition, the evaluation of both physiological and biochemical changes (e.g., chlorophyll content, proline content or shifts in the plant oxidative metabolism) could help in determining eventual stress-related responses to WD in these two crops.

Similarly to the yield parameters, WD treatment increased the concentration of total soluble proteins only in *L. culinaris* seeds. A higher protein amount after WD application has also been observed in rapeseed plants [33]. The present results are particularly relevant from a nutritional point of view, as leguminous crops are a central part of the human diet and could significantly compensate for the forecasted shortage of animal proteins [18]. However, it remains to be further investigated why *C. arietinum* and *P. vulgaris* did not respond as well. Also, in this case, the present results are at variance with those of [15], which found an increased protein content after WD foliar treatment, but without any fertigation.

The application of WD also enhanced the concentration of elements with high nutritional value, including Fe, Mg, Ca, and K. Deficiencies of these nutrients are highly prevalent among populations and could be alleviated via food fortification approaches. In this context, the production of seeds with a higher Fe content could critically contribute to reducing the number of subjects affected by iron deficiency, which is the micronutrient treated with the most concern worldwide regarding human nutrition [34,35]. Likewise, improved lentil seeds with higher K or chickpea seeds with higher Mg and Ca could contribute to reducing nutritional deficiencies related to diabetes, hypokalemia or osteoporosis [36,37]. Seeds from WD-treated *P. vulgaris* showed a decrease in Zn concentration, while this element was not affected in *C. arietinum* or in *L. culinaris*. Although Zn concentration in *P. vulgaris* was slightly lower compared to the ranges previously reported (10 $\mu g\ g^{-1}$ vs. 13 to 33.3 $\mu g\ g^{-1}$ [38,39]), the causes of this decrease in response to WD remain to be further investigated. It is worth noting that an increase in Na levels was observed in seeds from WD-treated plants. The concentration of this element both in control and WD-treated plants was below the values reported in the literature in the case of chickpea seeds (67–127 $\mu g\ g^{-1}$ vs. 188–285 $\mu g\ g^{-1}$ DW [40–42]) and lentil (64–124 $\mu g\ g^{-1}$ vs. 300–790 $\mu g\ g^{-1}$ DW [43,44]), although it was above that reported for

P. vulgaris (333–492 μg g^{-1} vs. 30–180 μg g^{-1}; [38,45]). Nevertheless, the concentration of Na in the diluted WD employed in this study (~15 μg L^{-1}) was far below the 0.5 g L^{-1} 'no risk' limit set by the FAO concerning salt concentrations in irrigation water [46]. To the best of our knowledge, how the application of WD enhances the concentration of elements such as Fe, Mg, Ca or Na in the seed remains unknown. A higher expression of auxin-responsive genes in WD-treated tomato plants was recently observed by [47]. In relation to this, WD could contain auxin-like substances that induce ATPase pumps in root tissues, thus favoring the entrance of cations, as previously shown in humic substances [48]. However, there is no experimental basis that has yet proven this hypothesis. Therefore, while initial results are encouraging, further studies are needed to dig into the processes influenced by WD that generate such differences in the mineral composition across the species tested, such as their absorption into the roots or their allocation in the seed. In parallel, future studies should also focus on the effects of WD on the plant-associated microbiota, as microorganisms critically contribute to maintaining plant growth and health in agricultural systems.

5. Conclusions

The present study showed the positive effect of WD treatment on the yield and nutritional composition of three legume species. Lentil plants showed an increased yield in terms of both pod and seed weight/number, as well as an improved concentration of total soluble proteins. Moreover, the treatment with WD differentially enhanced the content of mineral elements of high nutritional value in the seeds, such as Fe, K, Ca, and Mg. Overall, these results suggest that WD is an optimal candidate to sustainably boost sources of proteins and micronutrients in the human diet by generating high-yielding legume seeds with improved nutritional quality. Given the widespread impact of nutritional deficiencies on general health, the correction of these deficiencies through WD-based food fortification approaches should be considered. Lastly, and in line with the Farm to Fork strategy of the European Union, this study suggests that WD holds huge potential as a novel "green" alternative in the replacement of chemical fertilizers.

Author Contributions: Conceptualization: E.B., I.C., A.C., C.G. and S.L.; Methodology: E.B., P.C., I.C., C.G. and S.L.; Formal analysis and investigation: E.B., P.C., C.C., A.C., M.D. and L.P.; Writing—original draft preparation: P.C.; Writing—review and editing: E.B., P.C., I.C., C.C., A.C., M.D., C.G., S.L. and L.P.; Funding acquisition: S.L.; Resources: I.C., C.G. and L.P.; Supervision: I.C. and S.L. All authors have read and agreed to the published version of the manuscript.

Funding: This study was supported by the University of Florence (Fondi di Ateneo 2021) and by the University of Siena (project CLASS "Ceci, distillato di legno, biochar e intelligenza artificiale per un sistema agrifood smart e sostenibile" funded by Regione Toscana and Esperia Srl.).

Data Availability Statement: The data presented in this study are available on request from the corresponding author.

Acknowledgments: The authors want to thank the University of Florence (Fondi di Ateneo 2021) and the University of Siena (Project CLASS "Ceci, distillato di legno, biochar e intelligenza artificiale per un sistema agrifood smart e sostenibile" funded by Regione Toscana and Esperia Srl.) for the funding. The authors are also grateful to Del Colle Srl. enterprise for having kindly provided the legume seeds and the crop fields used for the experiment, as well as to Francesco Barbagli (BioDea and BioEsperia Srl.) for kindly providing the wood distillate. The authors are also grateful to Matteo De Filippis and Prandvera Guri for their contribution in the experimental work. The publication was made with the contribution of the researcher M.D., with a research contract co-funded by the European Union—PON Research and Innovation 2014–2020 in accordance with Article 24, paragraph 3a), of Law No. 240 of 30 December 2010, as amended, and Ministerial Decree No. 1062 of 10 August 2021.

Conflicts of Interest: The authors declare no conflict of interest.

References

1. World Health Organization. *The State of Food Security and Nutrition in the World 2020: Transforming Food Systems for Affordable Healthy Diets*; Food and Agriculture Organization of the United Nations: Rome, Italy, 2020; Volume 2020.
2. FAO Statistical Yearbook 2022. 2022. Available online: http://www.fao.org (accessed on 15 January 2023).
3. Grewal, A.; Abbey, L.; Gunupuru, L.R. Production, prospects and potential application of pyroligneous acid in agriculture. *J. Anal. Appl. Pyrolysis* **2018**, *135*, 152–159. [CrossRef]
4. Italian Ministerial Decree 6793. 18 July 2018. Available online: https://www.gazzettaufficiale.it/eli/id/2018/09/05/18A05693/sg (accessed on 20 June 2022).
5. Mathew, S.; Zakaria, Z.A. Pyroligneous acid—The smoky acidic liquid from plant biomass. *Appl. Microbiol. Biotechnol.* **2015**, *99*, 611–622. [CrossRef] [PubMed]
6. Terzopoulou, P.; Kamperidou, V. Chemical characterization of Wood and Bark biomass of the invasive species of Tree-of-heaven (*Ailanthus altissima* (Mill.) Swingle), focusing on its chemical composition horizontal variability assessment. *Wood Mater. Sci. Eng.* **2022**, *17*, 469–477. [CrossRef]
7. Agoncillo, E.S.; Alcate, V.; Philippines, O.M. Vegetable Seed Germination Enhancement Using Different Levels of Pyroligneous Acid (PA). *Glob. J. Biol. Agric.* **2018**, *2*, 14–18.
8. Vannini, A.; Moratelli, F.; Monaci, F.; Loppi, S. Effects of wood distillate and soy lecithin on the photosynthetic performance and growth of lettuce (*Lactuca sativa* L.). *SN Appl. Sci.* **2021**, *3*, 113. [CrossRef]
9. Vannini, A.; Fedeli, R.; Guarnieri, M.; Loppi, S. Foliar application of wood distillate alleviates ozone-induced damage in lettuce (*Lactuca sativa* L.). *Toxics* **2022**, *10*, 178. [CrossRef]
10. Fedeli, R.; Vannini, A.; Guarnieri, M.; Monaci, F.; Loppi, S. Bio-baseds solutions for agriculture: Foliar application of wood distillate alone and in combination with other plant-derived corroborants results in different effects on lettuce (*Lactuca sativa* L.). *Biology* **2022**, *11*, 404. [CrossRef]
11. Becagli, M.; Santin, M.; Cardelli, R. Co-application of wood distillate and biochar improves soil quality and plant growth in basil (*Ocimum basilicum*). *J. Plant Nutr. Soil Sci.* **2022**, *185*, 120–131. [CrossRef]
12. Ofoe, R.; Qin, D.; Gunupuru, L.R.; Thomas, R.H.; Abbey, L. Effect of pyroligneous acid on the productivity and nutritional quality of greenhouse tomato. *Plants* **2022**, *11*, 1650. [CrossRef]
13. Zulkarami, B.; Ashrafuzzaman, M.; Husni, M.O.; Ismail, M.R. Effect of pyroligneous acid on growth, yield and quality improvement of rockmelon in soilless culture. *Aust. J. Crop Sci.* **2011**, *5*, 1508–1514.
14. Simma, B.; Polthanee, A.; Goggi, A.S.; Siri, B.; Promkhambut, A.; Caragea, P.C. Wood vinegar seed priming improves yield and suppresses weeds in dryland direct-seeding rice under rainfed production. *Agron. Sustain. Dev.* **2017**, *37*, 1–9. [CrossRef]
15. Fedeli, R.; Vannini, A.; Celletti, S.; Maresca, V.; Munzi, S.; Cruz, C.; Alexandrov, D.; Guarnieri, M.; Loppi, S. Foliar application of wood distillate boosts plant yield and nutritional parameters of chickpea. *Ann. Appl. Biol.* **2023**, *182*, 57–64. [CrossRef]
16. Del Borghi, A.; Strazza, C.; Magrassi, F.; Taramasso, A.C.; Gallo, M. Life Cycle Assessment for eco-design of product–package systems in the food industry—The case of legumes. *Sustain. Prod. Consum.* **2018**, *13*, 24–36. [CrossRef]
17. Venter, C.S.; Van, E.E. More legumes for better overall health. *S. Afr. J. Clin. Nutr.* **2001**, *172*, 280.
18. Śmiglak-Krajewska, M.; Wojciechowska-Solis, J. Consumption preferences of pulses in the diet of polish people: Motives and barriers to replace animal protein with vegetable protein. *Nutrients* **2021**, *13*, 454. [CrossRef]
19. Considine, M.J.; Siddique, K.H.; Foyer, C.H. Nature's pulse power: Legumes, food security and climate change. *J. Exp. Bot.* **2017**, *68*, 1815–1818. [CrossRef]
20. Calles, T. The international year of pulses: What are they and why are they important. *Agric. Dev.* **2016**, *26*, 40–42.
21. Suter, M.; Connolly, J.; Finn, J.A.; Loges, R.; Kirwan, L.; Sebastià, M.T.; Lüscher, A. Nitrogen yield advantage from grass–legume mixtures is robust over a wide range of legume proportions and environmental conditions. *Glob. Chang. Biol.* **2015**, *21*, 2424–2438. [CrossRef]
22. Van Loon, M.P.; Deng, N.; Grassini, P.; Edreira, J.I.R.; Wolde-Meskel, E.; Baijukya, F.; Marou, H.; van Ittersum, M.K. Prospect for increasing grain legume crop production in East Africa. *Eur. J. Agron.* **2018**, *101*, 140–148. [CrossRef]
23. Sidhu, J.S.; Zafar, T.; Benyathiar, P.; Nasir, M. Production, processing, and nutritional profile of chickpeas and lentils. In *Dry Beans and Pulses: Production, Processing, and Nutrition*; John Wiley & Sons Ltd.: Hoboken, NJ, USA, 2022; pp. 383–407. [CrossRef]
24. Shrestha, S.; van't Hag, L.; Haritos, V.S.; Dhital, S. Lentil and Mungbean protein isolates: Processing, functional properties, and potential food applications. *Food Hydrocoll.* **2022**, *135*, 108142. [CrossRef]
25. Branlard, G.; Bancel, E. Protein extraction from cereal seeds. In *Plant Proteomics: Methods and Protocols*; Humana Press Inc.: Totowa, NJ, USA, 2007; pp. 15–25. [CrossRef]
26. Bradford, M.M. A rapid and sensitive method for the quantitation of microgram quantities of protein utilizing the principle of protein-dye binding. *Anal. Biochem.* **1976**, *72*, 248–254. [CrossRef] [PubMed]
27. Bettarini, I.; Colzi, I.; Coppi, A.; Falsini, S.; Echevarria, G.; Pazzagli, L.; Selvi, F.; Gonnelli, C. Unravelling soil and plant metal relationships in Albanian nickel hyperaccumulators in the genus *Odontarrhena* (syn. *Alyssum* sect. *Odontarrhena*, Brassicaceae). *Plant Soil* **2019**, *440*, 135–149. [CrossRef]
28. Bianchi, E.; Coppi, A.; Nucci, S.; Antal, A.; Berardi, C.; Coppini, E.; Fibbi, D.; Del Bubba, M.; Gonnelli, C.; Colzi, I. Closing the loop in a constructed wetland for the improvement of metal removal: The use of *Phragmites australis* biomass harvested from the system as biosorbent. *ESPR* **2021**, *28*, 11444–11453. [CrossRef] [PubMed]

29. Pangnakorn, U.; Watanasorn, S.; Kuntha, C.; Chuenchooklin, S. Application of wood vinegar to fermented liquid bio-fertilizer for organic agriculture on soybean. *Asian J. Food Agro-Ind.* **2009**, *2*, S189–S196.
30. Rogelio, R.M. Alternative growth ehancers for rice production: Usefulness of wood vinegar (PA) in irrigated rice (PSB rc18). *J. Biol. Agric. Healthc.* **2018**, *8*, 82–98.
31. Mu, J.; Yu, Z.M.; Wu, W.Q.; Wu, Q.L. Preliminary study of application effect of bamboo vinegar on vegetable growth. *For. Ecosyst.* **2006**, *8*, 43–47. [CrossRef]
32. Zhu, K.; Gu, S.; Liu, J.; Luo, T.; Khan, Z.; Zhang, K.; Hu, L. Wood vinegar as a complex growth regulator promotes the growth, yield, and quality of rapeseed. *Agronomy* **2021**, *11*, 510. [CrossRef]
33. Mao, K.; Li, S.; Li, B.; Wu, W.; Wei, C.; Yuan, S.; Niu, Y.; Du, H.; Zhang, L. Effect of wood vinegar on growth, yield and quality of upper leaves of flue-cured tobacco of Nanzheng. *Southwest China J. Agric. Sci.* **2019**, *32*, 645–652.
34. Mungkunkamchao, T.; Kesmala, T.; Pimratch, S.; Toomsan, B.; Jothityangkoon, D. Wood vinegar and fermented bioextracts: Natural products to enhance growth and yield of tomato (*Solanum lycopersicum* L.). *Sci. Hortic.* **2013**, *154*, 66–72. [CrossRef]
35. Guzmán-Maldonado, S.H.; Acosta-Gallegos, J.; Paredes-López, O. Protein and mineral content of a novel collection of wild and weedy common bean (*Phaseolus vulgaris* L). *J. Sci. Food Agric.* **2000**, *80*, 1874–1881. [CrossRef]
36. Venn, B.J.; Mann, J.I. Cereal grains, legumes and diabetes. *Eur. J. Clin. Nutr.* **2004**, *58*, 1443–1461. [CrossRef]
37. Rebello, C.J.; Greenway, F.L.; Finley, J.W. A review of the nutritional value of legumes and their effects on obesity and its related co-morbidities. *Obes. Rev.* **2014**, *15*, 392–407. [CrossRef]
38. Petropoulos, S.A.; Fernandes, Â.; Plexida, S.; Chrysargyris, A.; Tzortzakis, N.; Barreira, J.C.; Barros, L.; Ferreira, I.C. Biostimulants application alleviates water stress effects on yield and chemical composition of greenhouse green bean (*Phaseolus vulgaris* L.). *Agronomy* **2020**, *10*, 181. [CrossRef]
39. Gelin, J.R.; Forster, S.; Grafton, K.F.; McClean, P.E.; Rojas-Cifuentes, G.A. Analysis of seed zinc and other minerals in a recombinant inbred population of navy bean (*Phaseolus vugaris* L.). *Crop Sci.* **2007**, *47*, 1361–1366. [CrossRef]
40. Ray, H.; Bett, K.; Tar'an, B.; Vandenberg, A.; Thavarajah, D.; Warkentin, T. Mineral micronutrient content of cultivars of field pea, chickpea, common bean, and lentil grown in Saskatchewan, Canada. *Crop Sci.* **2014**, *54*, 1698–1708. [CrossRef]
41. Ereifej, K.I.; Al-Karaki, G.N.; Hammouri, M.K. Seed chemical composition of improved chickpea cultivars grown under semiarid Mediterranean conditions. *Int. J. Food Prop.* **2001**, *4*, 239–246. [CrossRef]
42. Daur, I.; Khan, I.A.; Jahangir, M. Nutritional quality of roasted and pressure-cooked chickpea compared to raw (*Cicer arietinum* L.) seeds. *Sarhad J. Agric.* **2008**, *24*, 117.
43. Zia-Ul-Haq, M.; Ahmad, S.; Shad, M.A.; Iqbal, S.; Qayum, M.; Ahmad, A.; Luthria, D.L.; Amarowicz, R. Compositional studies of lentil (*Lens culinaris* Medik.) cultivars commonly grown in Pakistan. *Pak. J. Bot.* **2011**, *43*, 1563–1567.
44. Sahi, S.T.; Ghazanfar, M.U.; Afzal, M.; Wakil, W.; Habib, A. Influence of inoculation with Ascochyta lentis on mineral contents (Na, Ca, Mg, Zn, Cu and Fe) of susceptible and resistant lines of lentil (*Lens culinaris* Medik.). *Pak. J. Bot.* **2010**, *42*, 375–382.
45. Paredes, M.; Becerra, V.; Tay, J. Inorganic nutritional composition of common bean (*Phaseolus vulgaris* L.) genotypes race Chile. *Chil. J. Agric. Res.* **2009**, *69*, 486–495. [CrossRef]
46. Brouwer, C.; Goffeau, A.; Heibloem, M. Chapter 7: Salty soils. In *Irrigation Water Management: Training Manual No. 1-Introduction to Irrigation*; FAO—Food and Agriculture Organization of the United Nations: Rome, Italy, 1985.
47. Ofoe, R.; Gunupuru, L.R.; Wang-Pruski, G.; Fofana, B.; Thomas, R.H.; Abbey, L. Seed priming with pyroligneous acid mitigates aluminum stress, and promotes tomato seed germination and seedling growth. *Plant Stress* **2022**, *4*, 100083. [CrossRef]
48. Canellas, L.P.; Teixeira Junior, L.R.L.; Dobbss, L.B.; Silva, C.A.; Medici, L.O.; Zandonadi, D.B.; Façanha, A.R. Humic acids crossinteractions with root and organic acids. *Ann. Appl. Biol.* **2008**, *153*, 157–166. [CrossRef]

environments

MDPI

Article

Zeolite and Winery Waste as Innovative By-Product for Vineyard Soil Management

Serena Doni [1,*], Grazia Masciandaro [1], Cristina Macci [1,*], Davide Manzi [1], Giovan Battista Mattii [2], Eleonora Cataldo [2], Maria Gispert [3], Francesca Vannucchi [1] and Eleonora Peruzzi [1]

1 National Research Council-Research Institute on Terrestrial Ecosystems (CNR-IRET), Via Moruzzi 1, 56124 Pisa, Italy; grazia.masciandaro@cnr.it (G.M.); davide.manzi@iret.cnr.it (D.M.); francesca.vannucchi@cnr.it (F.V.); eleonora.peruzzi@cnr.it (E.P.)

2 Department of Agriculture, Food, Environment and Forestry Sciences and Technologies (DAGRI), University of Florence, Viale delle Dee 30, 50019 Sesto Fiorentino, Italy; giovanbattista.mattii@unifi.it (G.B.M.)

3 Soil Science Unit, Department of Chemical Engineering, Agriculture and Food Technology, University of Girona, Campus Montilivi, C/Aurélia Capmany, 61, 17003 Girona, Spain; maria.gispert@udg.edu

* Correspondence: serena.doni@cnr.it (S.D.); cristina.macci@cnr.it (C.M.)

Abstract: In semiarid environments, vine cultivation is a land use with a high impact with regard to soil erosion, loss of organic matter and biodiversity, contamination, and compaction. In addition, the wine supply chain produces a considerable quantity of organic waste, which remains as residues in the ecosystem. Within this context, we developed a sustainable vine management system to improve the efficient use of fertilisers by applying a by-product derived from the composting of winery wastes and zeolite. We evaluated the effects of the zeolite-based compost on the chemical, physical, and biochemical soil properties of a productive vineyard. Four treatments were set up and monitored for about two years. These were as follows: (1) Commercial compost (COM); (2) Zeolite (Z); (3) 30% zeolite and 70% winery waste compost (30 ZEO); (4) 10% zeolite and 90% winery waste compost (10 ZEO). The results demonstrated that the ZEO treatments could be considered a win–win solution able to improve soil water content, nutrient retention, carbon sequestration, and biochemical activity while also recycling wastes. In particular, 10 ZEO seems to be the amendment that best combines an improvement in soil biochemical properties with gradual and constant nutrient availability, thus satisfying, without exceeding, soil and plant needs.

Keywords: soil management; winery wastes; zeolite; compost; carbon sequestration; nutrient availability

1. Introduction

Climate change, which has been taking place for several years, has reduced the water availability of Italian and European soils, thus altering the phenological phases of many crops. For grapevine, for example, there is an advance of 6–8 days for each degree of temperature increase during the growing season. In 2022, severe drought and extreme heat posed new threats to wine production. A Global Drought Observatory report indicated that almost two-thirds of the European territory was in a drought situation or on alert due to heat waves and deficient rainfall [1]. In fact, 2022 marked the worst drought in the last 500 years. The increase in anomalous grape harvests, with short and intense rainfall concentrated only in a few days a year, is also causing an increase in erosive phenomena, especially in the hillside vineyards with rows aligned down the maximum slope gradient [2].

Several studies have reported that European-wide vineyards tend to lose more soil than the proposed upper limit of the tolerable soil erosion rate of 1.4 t ha^{-1} $year^{-1}$ [3]. In addition, agronomic practices in vineyard soils, such as tillage, weed control via herbicides, and using mineral fertilizers at the expense of organic ones, has resulted in soil structure degradation and organic matter loss [4]. The decrease in organic matter, especially in intensive monocultures such as viticulture, is becoming a consolidated problem in the majority of Italian vineyards. At the European level, about 90% of soils have a low (0–2%) or medium (2–6%) organic carbon content. The lowest values of organic carbon are found above all in the soils of southern Europe, where 74% of the territory has less than 2% of organic carbon at a depth of 0–30 cm.

Tillage destroys soil structure, mixes surface horizons with a homogeneous distribution of organic matter in the worked layer and alters soil temperature. A loss in SOC of about 2.3 t ha^{-1} year has been estimated in the 0–17 cm horizon in the first five years of a vineyard's soil cultivation [5], while a decline in SOC of 15.7 t h^{-1} year has been measured in a mountain vineyard [6]. Additionally, using inorganic fertiliser as a significant soil nutrient replacement resulted in negative environmental repercussions, such as soil acidification and waterbody contamination [7]. In cases of soil acidification an inorganic liming treatment could be considered.

As reported by the International Organization of Vine and Wine (IOVW), Italy is the top wine producer, producing around 50 million hl during the 2021 campaign [8]. For this reason, an efficient management of vineyards in Italy is of paramount importance.

As is well known, the wine production process involves generating significant amounts of organic waste (stalks and grape pomace). Pomace represents about 18–25% of the grape mass that becomes a by-product during wine pressing and fermentation [9]. Similarly, stalks that make up the lignocellulosic skeleton of grapes are obtained after the destemming process and represent between 3 and 6% of the grapes [10].

Separation and exploitation of these by-products are critical imperatives of sustainability. Such organic wastes represent a resource rather than an annoying cost to manage. In Italy, some estimates indicate a potential availability of about 2.4 million tons of winemaking by-products; on average, three-quarters are unused.

The European Commission launched a public consultation on organic farming in 2020 in order to foresee 25% of land under organic cultivation by 2030. The commission budgeted EUR 40 million in 2021 to support organic farming, guarantee sustainable soil management, and promote the use of organic products so as to accomplish this goal, which is included in the European Green Deal. Many studies have been conducted in several countries to assess the potential of compost application in fertility restoration. These include various sources of organic residues such as cattle manure and cowpea haulm, animal manure, kitchen and kraal manure, cherry residue, cattle manure and sugarcane bagasse, and pig manure.

Recycling winery wastes as compost in agriculture can have several benefits. These comprise the conservation of limited and non-renewable rock phosphate utilized as chemical P fertilizer, improvement in soil nutrient content, structure stabilization of soil, reduction in soil erosion, and mitigation of climate change due to reduced GHG emissions from waste decomposition in open dumps. Furthermore, adding zeolite has proved to be an effective strategy to conserve nitrogen and mitigate gaseous emissions during composting of organic agricultural wastes [11].

Zeolites are crystalline, hydrated aluminosilicates that can swap their constituent elements and lose and acquire water reversibly without experiencing a significant structural change [12].

Previous studies have demonstrated that a 1–30% zeolite dosage during composting of various organic wastes, including winery wastes, enabled higher retention of nutrients, stimulation of enzymatic activity, and humification of organic matter to control composting (100% wastes) [13,14]. In addition, in zeolite-based composts a decrease in electrical conductivity and an increase in germination index were observed, thus indicating the improvement in the agronomic quality of the compost [15,16].

When added to soil, zeolite can operate as a long-lasting, nutrient-conserving amendment, minimising nutrient leaching and improving water usage efficiency while also promoting crop development [17,18]. Natural zeolites are particularly capable of absorbing NH_4^+ from solutions in a variety of environments and releasing it under proper conditions.

They exhibited a remarkable selectivity for cations with low ionic potential, such as K^+ and NH_4^+ [18]. It is important to note that the effectiveness of zeolites in nutrient retention can vary based on factors such as zeolite type, particle size, application rate, and soil conditions.

Within this framework, we started from the results of previous studies aimed at separately evaluating the effectiveness of zeolite and compost application in the improvement of agricultural soil quality. The objective of the present research was to examine the influence of the application of an innovative compost based on zeolite and winery wastes on vineyard soil quality and fertility in comparison with a commercial compost. At the same time, compost production, starting from wine processing waste, offers a solution to the problem of waste management by closing the company's production cycle.

2. Material and Methods

In spring 2019, the following treatments were applied in triplicate to a productive vineyard (*Vitis vinifera* L., Sangiovese cultivar): (1) Commercial compost (COM) at the rate 20 t ha^{-1} (6.6 t C ha^{-1}); (2) Zeolite (Z) at the rate 10 t ha^{-1}; (3) Compost based on 10% zeolite and 90% winery waste (10 ZEO) at the rate 30 t ha^{-1} (8.7 t C ha^{-1}); (4) Compost based on 30% zeolite and 70% winery waste (30 ZEO) at the rate 30 t ha^{-1} (7.8 t C ha^{-1}); (5) Control soil (C). The zeolite composts were obtained as described by Doni et al. [19].

The experimentation was set up in San Miniato (Pisa, Tuscany) in central Italy. The climate is typically Mediterranean and semiarid, with a mean annual precipitation of 859 mm and a mean annual temperature of 14.3 °C. The soil classification is Calcixerept [20] with a sandy clay loam texture (51.1% sand, 28.3% clay, and 20.6% silt) (USDA classification), an organic matter (OM) content of 1.8% (±0.2), a high level of carbonate (bivalve shells were widespread), and a slightly alkaline pH.

In this productive vineyard, the vine spacing is 2 m between rows and 0.8 m between plants. Before the beginning of the vegetative growth of the vine (March 2019), the vineyard was divided into fifteen sub-plots of about 0.15 ha each (8 m × 20 m), where the treatments were applied in triplicate (four treatments repeated three times). The sub-plots were chosen and considered to be independent true replicates, and a 2 m wide strip between them was considered to reduce any edge effect. The experimental site had a completely randomized block design, as shown in Figure 1.

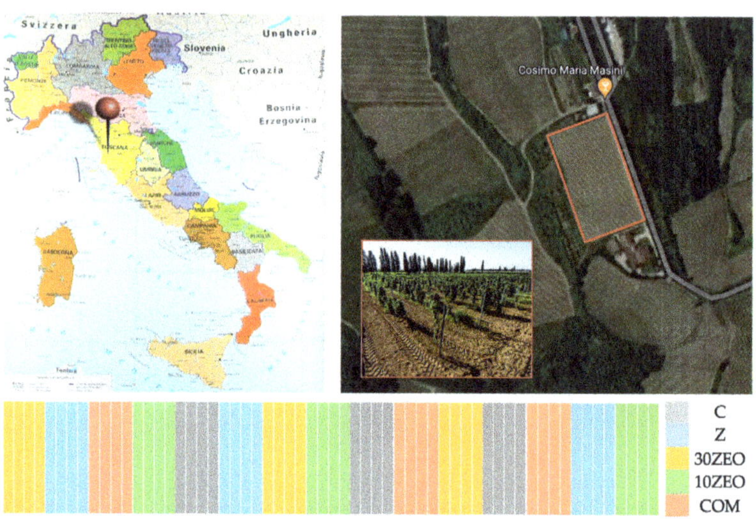

Figure 1. Location of the wine-growing area and experimental design of the treatments. C: control soil; Z: zeolite; 30 ZEO: 30% zeolite and 70% winery waste compost; 10 ZEO: 10% zeolite and 90% winery waste compost; COM: commercial compost.

The management system of the vineyard was characterized by commercial compost fertilization at a rate of 80 t/ha every three years. The selected vineyard received its last fertilization using commercial compost in winter 2016.

A manure spreader was used to distribute the compost-based treatments (COM, 30 ZEO, and 10 ZEO) over the entire soil surface. At the same time, zeolite with a granulometry of 1.0–2.5 mm was uniformly distributed through a centrifugal spreader capable of performing variable rates. All treatments were buried via harrowing to a depth of 30 cm.

The X-ray diffractometry (XRD) analysis of zeolite showed that it was almost 85% clinoptilolite, and the remainder comprised stilbite and heulandite. Table 1 reports the zeolite main characteristics. Table 2 shows the different compost characteristics.

Table 1. Chemical characteristics of zeolite (Z treatment).

	Zeolite-Clinoptilolite (Z)
SiO_2 %	68.3 ± 4.2
K_2O %	2.8 ± 0.31
Na_2O %	0.75 ± 0.07
Al_2O_3 %	12.3 ± 1.03
Fe_2O_3 %	1.3 ± 0.14
TiO_2 %	0.15 ± 0.03
CaO %	3.9 ± 0.25
MgO %	0.9 ± 0.05
Loss on ignition %	12.5 ± 0.95
Ration Si/Al %	5.1 ± 0.42
CEC $cmol_c\ kg^{-1}$	130 ± 21

CEC: Cation Exchange Capacity.

Table 2. Chemical and biological characteristics of compost treatments (COM, 10 ZEO, and 30 ZEO treatments).

		COM	**10 ZEO**	**30 ZEO**	**Italian Legislation 75/2010**	**EU Legislation 1009/2019**
pH		8.6 ± 0.05	8 ± 0.19	8.3 ± 0.16	6–8.8	
EC	$dS\ m^{-1}$	1.8 ± 0.01	0.4 ± 0.01	0.2 ± 0.02		
CEC	$cmolc\ kg^{-1}$	41 ± 2.1	44 ± 1.5	46 ± 1.5		
TOC	C %	33 ± 1.4	29 ± 0.6	26 ± 0.5	≥20	
TN	TN%	3.5 ± 0.06	1.55 ± 0.13	1.5 ± 0.08		
C/N		9.4	18.7	17.3	≤50	
HC	C%	6.1 ± 0.05	10.4 ± 0.07	11.8 ± 0.08	≥2.5	
TK	%	0.75 ± 0.005	0.8 ± 0.012	1.19 ± 0.019		
TP	%	0.13 ± 0.003	0.17 ± 0.004	0.14 ± 0.004		
Cu	$mgCu\ kg^{-1}$	49 ± 4	57 ± 6	44 ± 3	<230	<300
Salmonella	MPN/25 g	Absent	Absent	Absent	Absent	Absent
E. coli	CFU/g	<1000	<1000	<1000	<1000	<1000
GI	%	115 ± 1.5	126 ± 8.1	142 ± 6.1	>60%	

EC: electrical conductivity; CEC: Cation Exchange Capacity; TOC: Total Organic Carbon; TN: total nitrogen; HC: humic carbon; TK: total potassium; TP: total phosphorus; GI: Germination Index.

Immediately after treatment application (Spring 2019, T0) and after the grape harvesting in 2019 (six months after treatments application, October 2019, T1), 2020 (18 months after treatments application, October 2020, T2), and 2021 (30 months after treatments application, October 2021, T3), three composite soil samples were taken from each sub-plot between vine rows at the 0–30 cm layer. Each soil sample was a composite of three sub-samples collected in a homogenous area. The three sub-samples were mixed, homogenized, and sieved (2 mm); a part of each of the composite soil samples was stored at four °C for inorganic N quantification, and another part was air-dried for the other analyses. The chemical and biochemical results are the average of the nine soil samples (3 composite soil samples × 3 replicates) for each treatment (9 soil samples × 5 treatments) for each sampling time (45 × 4 sampling campaigns) for a total of 180 soil samples.

2.1. Chemical Parameters

Electrical conductivity (EC) and pH were determined in water extracts using selective electrodes Conmet 2 (Hanna Instruments, Woonsocket, Rhode Island, USA) and pH electrode InLab routine pro (Sevenmulti, Mettler Toledo, Greifensee, Switzerland). Total Organic Carbon (TOC) and total nitrogen (TN) were measured via dry combustion using FlashSmart Elemental Analyzer (Thermo Fisher Scientific, Waltham, MA, USA) according to the manufacturer's instructions. Total phosphorus (TP) and potassium (TK) were extracted via nitric-perchloric acid digestion (HNO_3:$HClO_4$, 5:2) in a microwave Ethos 1 (Milestone srl, Bergamo, Italy). Available phosphorus (AP) was extracted using the Olsen method [21]. TP and AP were measured using the method reported by Murphy and Riley (1962) using the Spectophotometer-Unicam UV 500 (Thermo Fisher Scientific, Waltham, MA, USA). Available potassium (AK) was extracted using ammonium acetate solution [22]. TK and AK were determined using 5900 ICP-OES (Agilent, Santa Clara, CA, USA). Sodium pyrophosphate (0.1 M, pH 11) at 60 °C for 24 h under shaking at 200 oscillation min^{-1} was used to extract Total Humic Carbon (THC). The THC extract was separated into humic (HA) and fulvic (FA) acids via the addition of H_2SO_4; the extract was kept overnight at 4 °C and then the flocculent (HA) and the supernatant (FA) were centrifuged. THC and FA were determined according to the method proposed by Jakubus and Spychalski [23] in which $KMnO_4$ oxidized carbon. The HA was calculated by subtracting FA from THC. The Cation Exchange Capacity (CEC) of the soils was determined via the Sumner and Miller method [24] using barium chloride (pH 8.1).

2.2. Biochemical Parameters

Total β-glucosidase (Glu), phosphatase (Pho), arylsulphatase (AS), and butyrate esterase (BE) activities were tested via the method of Marx et al. [25] and Vepsäläinen et al. [26] using 4-mthylumbelliferyl β-glucosidase, 4-methylumbelliferyl phosphate, 4-methylumbelliferyl sulphate, and 4-methylumbelliferyl butyrate, respectively, as substrates. Fluorescence (excitation 360 nm; emission 450 nm) of the product 4-Methylumbelliferone was measured using an Infinite F200 pro plate reader fluorimeter (Tecan, Männedorf, Zürich, Switzerland) after 0, 30, 60, 120, and 180 min of incubation at 30 °C.

2.3. Physical Parameters

According to the manufacturer's instructions, the wet stability of aggregates (WSA) was measured at plot level on aggregates extracted from the air-dry sieved soil (Eijkelkamp, Soil & Water, Giesbeek, The Netherlands). Briefly, 4 g of 1–2 mm air-dry aggregates were transferred to a sieve with 250 µm openings and rewetted using a vaporizer. The sieve was moved up and down for 3 min (34 cycles min^{-1}; stroke length 13 mm) in aluminum cans filled with 100 mL of water. The wet stability of aggregates was calculated as the stable soil aggregate fraction remaining on the sieve (after oven-drying at 105 °C for 24 h).

A pressure plate was used to determine field capacity and wilting point at a pF of 2.54 and 4.2, respectively. Available water (AW) was calculated as the difference between these two features [27].

2.4. Statistical Analysis

Prism 10.1 for macOS (GraphPad Inc., Dotmatics, Boston, MA, USA) was used for the statistical analysis. Analysis of variance was used to reveal differences between treatments and sampling times (one-way ANOVA). Fisher's test, which calculates the least significant differences at $p < 0.05$, was used to compare the means. Furthermore, potential patterns or clusters between treatments and physical–chemical and biochemical soil parameters were determined using principal component analysis (PCA). All raw data were log-transformed before being normalized, thereby reducing data heterogeneity. The PCs were only to be interpreted when component loadings greater than 0.60 were present.

3. Results

3.1. Chemical Parameters

The chemical properties of the soils treated using zeolite-based compost (ZEO treatments) compared to the control soil (C treatment) are reported in Table 3. The treatments did not influence the pH of the vineyard soil, and a slight increase in this parameter was observed in all soil samples over time. The CEC of the amended soils, especially 10 ZEO, was significantly higher at each sampling time than that of the unamended vineyard soil (C treatment).

Table 3. The effect of treatments (C, Z, 10 ZEO, 30 ZEO, and COM) on soil pH, Cation Exchange Capacity (CEC), total (TP and TK), and available (AP and AK) P and K of the vineyard soil immediately after treatments (T0), after six months (T1), 18 months (T2), and 30 months (T3).

	pH				CEC cmolc kg^{-1}			
	T0	**T1**	**T2**	**T3**	**T0**	**T1**	**T2**	**T3**
C	7.57 ± 0.37 aB	7.65 ± 0.18 aB	7.62 ± 0.02 aB	7.86 ± 0.46 aA	12.9 ± 0.97 cA	12.3 ± 0.91 bA	12.3 ± 0.92 cA	13.4 ± 0.43 cA
Z	7.70 ± 0.22 aAB	7.60 ± 0.37 aB	7.66 ± 0.25 aB	7.82 ± 0.27 aA	14.5 ± 0.70 bA	15.2 ± 0.82 aA	15.1 ± 1.13 bA	15.0 ± 1.10 bA
10 ZEO	7.57 ± 0.48 aB	7.51 ± 0.48 aB	7.68 ± 0.46 aAB	7.89 ± 0.25 aA	16.6 ± 0.72 aA	16.6 ± 0.98 aA	17.0 ± 1.27 aA	18.0 ± 0.40 aA
30 ZEO	7.66 ± 0.20 aA	7.75 ± 0.48 aA	7.77 ± 0.45 aA	7.79 ± 0.26 aA	15.0 ± 0.24 bA	15.6 ± 1.13 aA	15.7 ± 1.18 bA	16.1 ± 1.30 abA
COM	7.55 ± 0.20 aB	7.71 ± 0.26 aAB	7.66 ± 0.38 aB	7.89 ± 0.34 aA	17.0 ± 0.82 aA	16.7 ± 1.27 aA	16.1 ± 1.50 aA	16.7 ± 1.40 aA
	TP mgP kg^{-1}				**TK mgk kg^{-1}**			
	T0	**T1**	**T2**	**T3**	**T0**	**T1**	**T2**	**T3**
C	550 ± 27.5 bA	485 ± 69.3 cA	540 ± 25.6 bA	540 ± 24.6 Ac	5345 ± 45.4 bA	5333 ± 244 cdA	4152 ± 146 dB	4307 ± 159 bB
Z	532 ± 53.2 bAB	502 ± 41.8 Bc	522 ± 26.1 bB	582 ± 29.1 bA	5315 ± 243 bcB	5940 ± 48.6 bA	5339 ± 245 bB	5343 ± 245 aB
10 ZEO	626 ± 32.8 aB	714 ± 44.6 aA	653 ± 46.3 aAB	628 ± 49.7 abAB	6210 ± 318 aA	6242 ± 153 aA	6046 ± 91.3 aA	5418 ± 226 aB
30 ZEO	656 ± 29.9 aAB	619 ± 30.9 bB	597 ± 43.6 abB	697 ± 31.3 aA	5059 ± 221 cB	5553 ± 77.6 cA	5827 ± 285 abA	5553 ± 221 aA
COM	651 ± 32.5 aA	539 ± 27.0 cB	566 ± 28.3 bB	565 ± 28.2 bcB	5582 ± 265 bA	5158 ± 169 dAB	4934 ± 211 bcBC	4573 ± 181 bC
	AP mgP kg^{-1}				**AK mgP kg^{-1}**			
	T0	**T1**	**T2**	**T3**	**T0**	**T1**	**T2**	**T3**
C	3.28 ± 0.15 dA	3.45 ± 0.14 cA	3.64 ± 0.18 dA	3.70 ± 0.19 bA	242 ± 11.1 bcB	291 ± 10.5 bA	225 ± 11.3 bBC	198 ± 9.9 abC
Z	3.79 ± 0.16 cB	4.02 ± 0.17 bAB	4.30 ± 0.10 cA	3.39 ± 0.19 bB	220 ± 12.0 cB	248 ± 8.4 cA	196 ± 9.8 cBC	178 ± 12.9 bC
10 ZEO	5.52 ± 0.24 bA	3.56 ± 0.20 cB	3.63 ± 0.2 dB	3.64 ± 0.18 bB	249 ± 10.5 bB	284 ± 8.2 bA	256 ± 12.8 aB	178 ± 8.9 bC
30 ZEO	5.00 ± 0.21 bA	4.83 ± 0.15 aA	4.75 ± 0.15 bA	4.28 ± 0.16 aB	259 ± 12.0 bA	282 ± 14.1 bA	254 ± 12.7 aA	184 ± 9.2 bB
COM	6.84 ± 0.09 aA	5.00 ± 0.11 aB	5.24 ± 0.20 aB	3.76 ± 0.16 bC	324 ± 8.2 aA	346 ± 12.3 aA	274 ± 13.7 aB	204 ± 7.7 aC

Mean ± standard error. Lowercase letters indicate significantly different values at $p < 0.05$ between the treatments at each sampling time. Capital letters indicate significantly different values at $p < 0.05$ between the sampling times in the same treatment.

All compost-based treatments (COM, 10 ZEO, and 30 ZEO), particularly COM, enhanced soil electrical conductivity (EC) in comparison with the C soil at the initial sampling time (T0), whereas the Z treatment maintained it. This parameter showed a reduction over time in all treated soils, while it increased in the untreated control (C) soil (Figure 2).

Both 10 ZEO and 30 ZEO resulted in an increase in soil organic carbon content (TOC) compared to the C soil at each sampling time, and this increase was more pronounced in 10 ZEO at the final sampling (T3). As expected, an increase in TOC in the COM treatment compared to the C soil was also measured at T0 sampling time. However, during the T3 sampling campaign, there was no significant difference in the TOC between the COM treatment and the C soil (Figure 3).

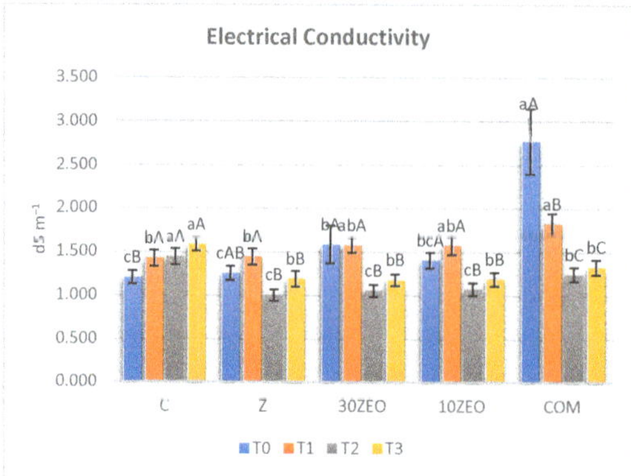

Figure 2. The effect of treatments (C, Z, 10 ZEO, 30 ZEO, and COM) on the electrical conductivity of the vineyard soil immediately after treatment (T0) and after 6 months (T1), 18 months (T2), and 30 months (T3). Mean ± standard error. Lowercase letters indicate significantly different values at $p < 0.05$ between the treatments at each sampling time. Capital letters indicate significantly different values at $p < 0.05$ between the sampling times in the same treatment.

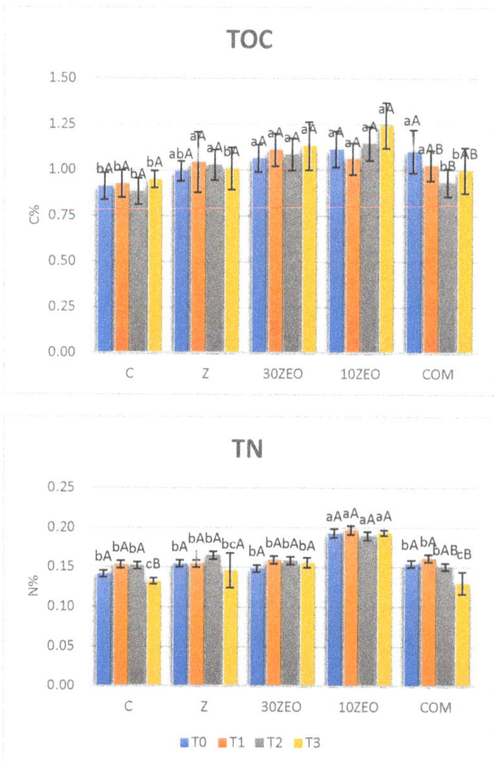

Figure 3. The effect of treatments (C, Z, 10 ZEO, 30 ZEO, and COM) on soil Total Organic Carbon (TOC) and total nitrogen (TN) of the vineyard soil immediately after treatment (T0) and after 6 months (T1), 18 months (T2), and 30 months (T3). Mean ± standard error. Lowercase letters indicate significantly different values at $p < 0.05$ between the treatments at each sampling time. Capital letters indicate significantly different values at $p < 0.05$ between the sampling times in the same treatment.

In a similar way, soil nitrogen (TN) was significantly affected by the treatments. An increase in this parameter in the Z and ZEO treatments, especially 10 ZEO, in comparison with the C soil was observed at the final sampling (T3) (10%, 47%, and 18% more TN than the C soil in Z, 10 ZEO, and 30 ZEO, respectively) (Figure 3). However, the TN content in the COM-treated soil decreased over time, and a significantly higher value was found in this treatment compared to the C soil only at the initial sampling time (T0).

Generally, the content of TP and TK, two of the essential plant nutrients, was higher in all the compost-treated soils (COM, 10 ZEO, and 30 ZEO), especially in 10 ZEO, compared with the C soil (Table 3). At the same time, the treatments containing zeolite (10 ZEO, 30 ZEO, and Z) decreased the available phosphorus (AP) and potassium (AK) compared to the COM treatment at T3 sampling, thus indicating the increase in soil retention ability of these elements.

Both 10 ZEO and 30 ZEO treatments significantly increased total humus (TH) compared to the control soil at each sampling time. In contrast, significantly higher values of this parameter in the COM treatment were observed only at the T0 sampling. In the COM treatment there was a decreasing trend, especially in fulvic acids, while the ZEO treatments maintained total humus over time and increased their humic acid fraction (Figure 4).

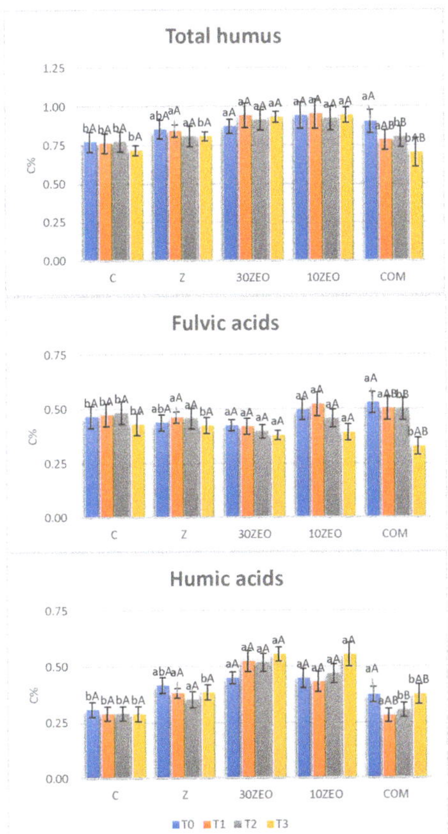

Figure 4. The effect of treatments (C, Z, 10 ZEO, 30 ZEO, and COM) on total humus, fulvic acids, and humic acids of the vineyard soil immediately after treatment (T0) and after 6 months (T1), 18 months (T2), and 30 months (T3). Mean ± standard error. Lowercase letters indicate significantly different values at $p < 0.05$ between the treatments at each sampling time. Capital letters indicate significantly different values at $p < 0.05$ between the sampling times in the same treatment.

3.2. Biochemical Parameters

Butyrate esterase and hydrolytic soil enzyme activities were determined at each sampling time (Figure 5).

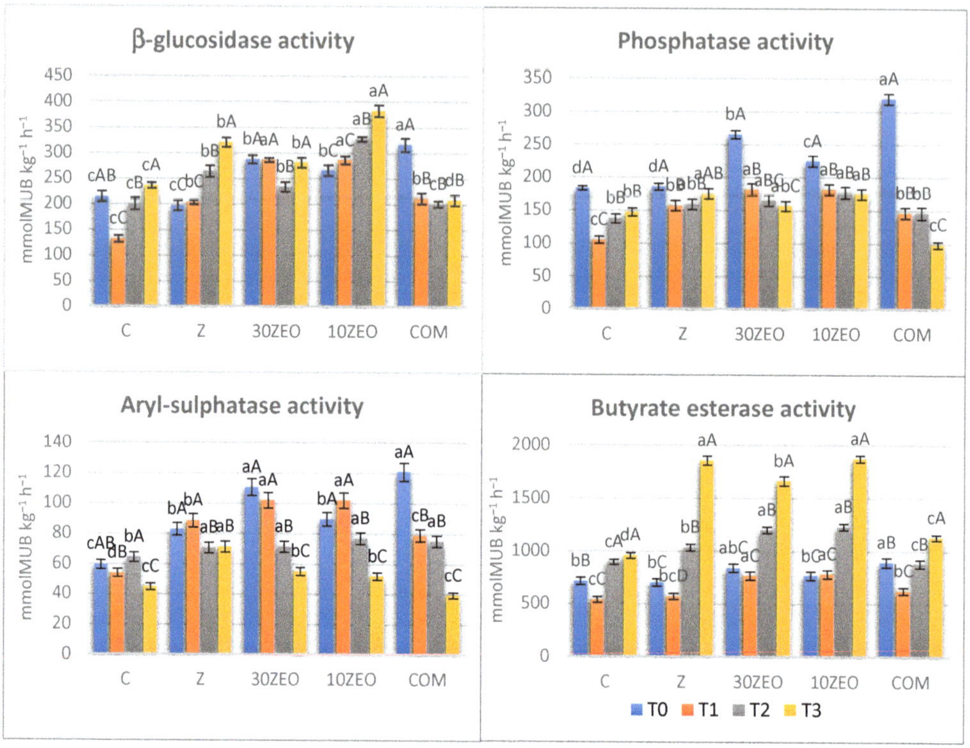

Figure 5. The effect of treatments (C, Z, 10 ZEO, 30 ZEO, and COM) on soil enzyme activities of the vineyard soil immediately after treatment (T0) and after 6 months (T1), 18 months (T2), and 30 months (T3). Mean ± standard error. Lowercase letters indicate significantly different values at $p < 0.05$ between the treatments at each sampling time. Capital letters indicate significantly different values at $p < 0.05$ between the sampling times in the same treatment.

10 ZEO, 30 ZEO, and Z treatments resulted in higher values of butyrate esterase activity (BE) at harvest time in the second year (T3). These treatments increased this activity by 74%, 96% and 94%, respectively, compared to the C soil. At the same sampling time, this activity increased by about 20% in the COM treatment.

The treatments influenced the activity of b-glucosidase (Glu). At the initial sampling time (T0), we observed a higher Glu activity in the COM treatment than in the other treatments (an increase of 48% compared to the C soil). In comparison, this treatment showed a lower activity in the second year of the study (T3) (11% lower than the C soil).

A similar trend was also shown by phosphatase (Pho) and aryl-sulphatase (AS) activities, which, at T3, were lower in the COM treatment than in the C soil (reductions of 33 and 13%, respectively). Both Pho and AS showed a decreasing trend in all treatments. However, higher values regarding the C soil were observed at each sampling time in all zeolite-based treatments (10 ZEO, 30 ZEO, and Z), especially in the Z treatment (increase of 18% Pho and 58% AS compared to the C soil).

3.3. Physical Parameters

The soil available water (AW) values were about 13.5% in the C soil at each sampling time (Table 4). With the ZEO treatments, both 10 ZEO and 30 ZEO, an increase in AW of about 1% was measured at the different sampling times (T0, T1, T2, and T3).

Table 4. The effect of treatments (C, Z, 10 ZEO, 30 ZEO, and COM) on soil wet aggregate stability (WSA) and available water (AW) of the vineyard soil immediately after treatment (T0) and after 6 months (T1), 18 months (T2), and 30 months (T3).

	WAS %				AW %			
	T0	T1	T2	T3	T0	T1	T2	T3
C	17.1 ± 0.86 aA	16.37 ± 0.82 aA	14.2 ± 1.43 aA	15.1 ± 0.75 aA	13.5 ± 0.58 aA	13.7 ± 0.69 aA	13.55 ± 0.48 aA	13.3 ± 0.66 aA
Z	30.7 ± 1.92 aB	29.0 ± 1.45 aC	27.8 ± 1.35 aB	28.1 ± 1.40 aC	14.2 ± 0.61 aAB	13.9 ± 0.70 aA	14.2 ± 1.42 aAB	14.6 ± 0.73 aAB
10 ZEO	27.4 ± 1.83 aAB	24.1 ± 1.18 aB	22.8 ± 1.14 AB	23.2 ± 1.15 aB	14.9 ± 0.74 aB	14.8 ± 1.24 aA	15.1 ± 1.06 aB	14.8 ± 0.54 aB
30 ZEO	23.8 ± 2.65 aB	23.5 ± 1.18 aB	24.3 ± 1.22 aAB	25.5 ± 1.27 aB	14.9 ± 0.74 aB	14.7 ± 1.47 aA	15.0 ± 0.75 aB	15.1 ± 0.65 aB
COM	21.6 ± 1.08 aA	22.7 ± 1.14 aB	19.4 ± 0.97 aAB	18.8 ± 0.94 aA	14.1 ± 0.70 aAB	14.6 ± 0.58 aA	14.8 ± 0.74 aAB	14.4 ± 0.66 aAB

Mean ± standard error. Lowercase letters indicate significantly different values at $p < 0.05$ between the treatments at each sampling time. Capital letters indicate significantly different values at $p < 0.05$ between the sampling times in the same treatment.

A greater wet aggregate stability (WAS) was noticed in ZEO with respect to the C and COM soils. In general, the 10 ZEO and 30 ZEO had intermediate values. All treatment effects were maintained over time (Table 4).

4. Discussion

The rise in electrical conductivity (EC) after the COM treatment is of concern because it can cause adverse effects on vine performance and grape composition [28]. However, *Vitis vinifera* L. (cv. Cabernet Sauvignon) is a crop considered moderately sensitive to salt [29], with a soil saturation extract threshold of EC at 25 °C of 2.6 dS m^{-1} for *Vitis vinifera* L. (cv. Sultana) [30].

However, the drop in EC observed in the ZEO and Z treatments has already been reported in a composting process using 5% and 10% zeolite [31] and in a study on the effects of soil additives, such as zeolites, on soil properties [32,33]. The high Cation Exchange Capacity (CEC) and molecular sieve structure of zeolite can enable ions to be exchanged within the environment, thereby lowering the EC of the soil [33].

Similarly, TOC showed different behavior in the ZEO and COM treatments, with an increase over time in ZEO and an increase immediately after treatment, followed by a decrease in COM. This trend indicates a priming effect on short-term SOC mineralization in the COM treatment, which modifies the microbial breakdown of SOC in response to carbon inputs.

Our study agrees with the results of previous studies where the compost treatment greatly stimulated bacteria to degrade organic carbon from the compost [34], and this is regarded as an essential part of the global carbon cycle, which may have a negative impact on the ability of soils to sequester carbon [35,36]. However, the soil can retain the added carbon in ZEO treatments, making zeolite a suitable material for carbon capture. It is reported that the interaction between minerals and organic matter can determine the preservation of organic carbon by limiting decomposer organisms from accessing organic materials [37,38]. In our specific case, composted organic matter can be protected by zeolite and can last longer in forms resistant to degradation (humic acids, for example). It is commonly known that the substantial specific surface area and CEC of zeolites enables them to absorb more organic carbon and have a high stabilizing capability [33,39,40]. According to our findings, adding natural zeolite to the soil in combination with organic matter, such as plant leftovers, enhanced carbon sequestration by increasing the soil's organic carbon content [41,42].

An assessment of the impact of the various treatments on the humification and mineralization processes of SOM were obtained by measuring the concentrations of total humus and its humic (HAs) and fulvic (FAs) acid fractions. A decrease in total humus was observed in the C, Z, and COM treatments over time, and in these treatments the content of FAs was generally higher than HAs, indicating a low humification rate [43].

In accordance with our results, a short-term study conducted in pots with poor sandy soil amended with zeolite and planted with *Lolium multiflorum* also discovered a decrease in humic compounds [44]. This drop was attributed to the ability of zeolite to improve the microbiological soil conditions by increasing nutrient and water retention, which altered the delicate balance between C sequestration and decomposition in soil. In addition, several studies have indicated that SOM turnover is either lowered in nutrient-deficient soils or increased after adding mineral fertilizers [45,46]. These considerations implied that while the ZEO treatments promoted the establishment of humification processes, as seen by the more pronounced increase in Has compared to FAs and the increase in total humus, the COM treatment might accelerate the microbial mineralization of TOC.

An upward trend in total N accompanied the increase in TOC and humic compounds in the ZEO-treated soils. In the COM treatment, ammonium is released during the mineralization of organic materials and is quickly transformed into nitrate, which is readily leached [47]. However, in ZEO treatments, there may be a significant amount of nitrogen from the added organic matter captured as ammonium by the zeolite [48]. Similarly, in the Z treatment, the increase in total N can be explained by the fact that the inorganic N can be released from soil organic N pools through physical and chemical interactions within the soil matrix and captured by zeolite up until plant absorption [49].

Zeolites have massive canals that can absorb cations like ammonium, but bacteria, especially nitrifying bacteria, cannot enter the zeolite canals [50]. As a result, ammonium, present in organic fertilisers or soil, can be selectively absorbed into zeolite, becoming unavailable to nitrify bacteria present in well-aerated sandy soils [8]. Additionally, ZEO treatments appear to prevent the volatilization of nitrogen (N) from soil because the zeolite absorbs N before releasing it. In this way, zeolite-based compost fertiliser can function as a slow-release fertiliser to provide N to the vine progressively.

Similarly, retention of other nutrients, such as P and K, could be equally crucial for the next season's plant growth. The Z and ZEO treatments showed a statistically significant increase in total K and a decrease in available K in comparison to the C and COM treatments. The adsorption of this cation is typical of clinoptilolite [51], the main component of the zeolite used in this study.

The trend of total P was comparable to that of total K. It may be attributable to the high nonspecific sorption of P, such as the electrostatic attraction of the negatively charged phosphate ions to the co-adsorbed cations (Na^+, Ca^{2+}, K^+, and Mg^{2+}) on natural zeolite [52]. Even if this process can be pH-dependent (the higher the pH, the lower the retention), a favorable electrostatic attraction mechanism in the P sorption by zeolite has been also proposed in the pH range 3–10 [53] and in the range of pH 4–8 [54]. However, other authors reported that electrostatic forces are high in acidic conditions when the binding sites are protonated, while a typical adsorption reaction that proceeds chemical interaction involving ligand exchange dominates at an alkaline pH [55].

In agreement with our results, although negative charges make PO_4^{3-} adsorption onto zeolite mineral lattices unfavorable, an increase in soil P content with increasing zeolite doses has previously been described [56–58]. Zeolites can also interact with phosphate ions through the filter effect of their three-dimensional structure. Clinoptilolite, the zeolite we used, shows a three-dimensional structure with the dimension of pores and channels compatible with many ionic substances.

All the treatments with organic matter and zeolite significantly increased the soil CEC. According to several studies, zeolite plays a pivotal role in improving the soil's physical structure. Further, through the modification of CEC, it can decrease nitrogen leaching, increase water retention capacity, and extend the persistence of nutrients in the root zone [59,60].

The beneficial effect of both ZEO treatments in improving the physical properties of vineyard soils was mainly due to the synergic effect of zeolite and organic matter. On the one hand, it is recognised that zeolite can reduce bulk density and enhance overall porosity, increasing soil's available water content [61]. On the other hand, the general positive effect of organic matter application in improving soil physical properties is recognized [62]. Notably, the improvement of physical properties in ZEO treatments is maintained during this time.

Improving the chemical–physical soil properties in zeolite-treated soils influenced the microbial biomass activity. From the perspective of soil quality, the change in microbial activity in soil is relevant since it carries out many critical functions in the soil ecosystem [63]. Microorganisms are fundamental in biogeochemical cycles and are essential to soil functions, such as soil organic matter decomposition and nutrient cycling, and, thus, provide crucial ecosystem services.

Butyrate esterase activity (BE), which has been used to evaluate overall soil microbial activity [64], increased in all treatments, especially in the presence of zeolite (10 ZEO, 30 ZEO, and Z), with respect to the C soil. This was due to the microbial biomass incorporated into the added compost and the positive influence of zeolite in improving the microenvironment for soil microbial activity. In addition, due to its porous structure, zeolite can provide physical protection and stabilize soil enzymes [65]. As expected, the COM treatment also enhanced the microbiological and biochemical soil conditions since the amount and activity of the soil microbial biomass are directly correlated with the C input [66,67]. The higher C content in the compost treatments (10 ZEO, 30 ZEO, and COM) increased microbial growth, which stimulated the utilization and assimilation of nutrients (i.e., N and P) via enzyme production [68]. The ability of soils to hydrolyse low molecular weight carbohydrates, represented by the β-glucosidase enzyme [69]; organic P compounds, represented by the phosphatase enzyme [70]; and organic S compounds, represented by the aryl-sulphatase activity [71], was monitored. These soil hydrolytic enzyme activities increased in all the treatments at T0 sampling time in comparison with the control soil. However, the rise in enzymatic activities, especially in COM treatment, did not persist over time. At T2, they diminished, most likely due to the reduction in their substrates probably following mineralization or humification processes of the organic substances contributed by the compost [72].

Principal Component Analysis (PCA)

The relationships between soil treatments, sampling times, and chemical–physical and biochemical soil properties were investigated using PCA analysis (Figure 6, Table 5). Two principal components were extracted. The PCs were rotated using the varimax approach, reducing the dataset dimension to two components; these explain 65.66% of the data variation, with PC1 explaining 40.57% and PC2 explaining 25.07% of the total variance. The indicators with factor loadings greater than 0.6 are CEC, TOC, TN, TP, TK TH, HA, GLU, WAS, and AW in PC1 and EC, pH, Kav, Pav, FA, BE, and AS in PC 2.

At each sampling time, ZEO-treated soils showed a shift towards positive values of the PC1 to the control soil. This principal component separated the ZEO treatments (positive values of PC1) from the COM and C treatments (negative values of PC1). This pattern indicates that there is a strong correlation between the ZEO treatments, in particular 10 ZEO, and the trend of the carbon biogeochemical cycle towards humification. This is shown by the positive correlation between the organic carbon content (TOC), even in the stabile form (HAs), and the C-linked activity (GLU) on the PC1. The positive correlation between soil water retention capacity (WRC) and Glu on this PC also confirmed the dominant role of soil moisture in biological functions [73]. This pattern additionally highlighted how the treatments, especially ZEO, can retain nutrients N, P, and K by raising CEC (parameters with significant positive factor loadings on the PC1).

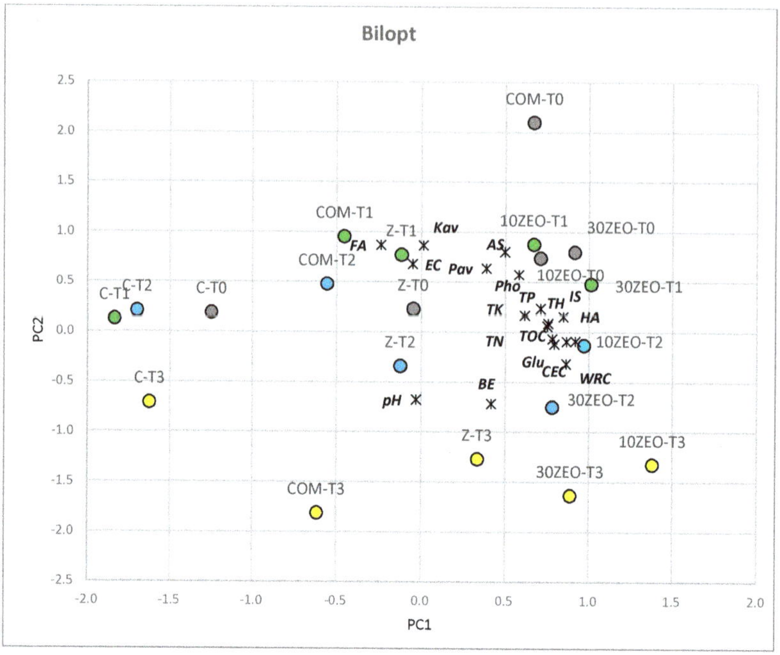

Figure 6. Biplot of factor scores and loadings in each treatment (C, Z, 30 ZEO, 10 ZEO, and COM) at T0 (grey), T1 (green), T2 (blue), and T3 (yellow) sampling times. CEC: Cation Exchange Capacity; EC: electrical conductivity; TOC: Total Organic Carbon; TN: total nitrogen; TP: total phosphorus; AP: available phosphorus; TK: total potassium; AK: available potassium; TH: total humus; FA: fulvic acids; HA: humic acids; Glu: b-glucosidase activity; Pho: phosphatase activity; BE: butyrate esterase activity; AS: aryl sulphatase activity; WAS: wet aggregate stability; AW: available water.

Table 5. The first two principal components (PCs) and component loadings were obtained by considering all treatments (C, Z, 30 ZEO, 10 ZEO, and COM) and all sampling times (T0, T1, T2, and T3).

	PC 1	PC 2
CEC	0.778 *	−0.067
EC	−0.051	0.679 *
pH	−0.030	−0.676 *
TOC	0.918 *	−0.091
TN	0.614 *	0.167
TP	0.753 *	0.086
AP	0.389	0.639 *
TK	0.709 *	0.236
AK	0.013	0.865 *
TH	0.845 *	0.153
FA	−0.238	0.868 *
HA	0.861 *	−0.317
Glu	0.789 *	−0.120

Table 5. *Cont.*

	PC 1	PC 2
Pho	0.581	0.574
BE	0.418	−0.716 *
AS	0.497	0.803 *
WAS	0.748 *	0.059
AW	0.863 *	−0.094
Explained variance (%)	40.58	25.08

CEC: Cation Exchange Capacity; EC: electrical conductivity; TOC: Total Organic Carbon; TN: total nitrogen; TP: total phosphorus. AP: available phosphorus; TK: total potassium; AK: available potassium; TH: total humus; FA: fulvic acids; HA: humic acids; Glu: b-glucosidase activity; Pho: phosphatase activity; BE: butyrate esterase activity; AS: aryl sulphatase activity; WAS: wet aggregate stability; AW: available water. *Variables with component loadings used to interpret the PCs; threshold level: 0.60.

The second PC (PC2) separated the sampling time of each treatment. For each soil sampling treatment, the initial time (T0) is shifted towards positive values of the PC2 with respect to the final time (T3). The T0 and T3 of the COM treatment are more distant on this PC, indicating the lower capacity of this treatment in modifying the chemical–physical properties of the treated soils.

5. Conclusions

The grapevine is one of the most economically valuable crops in the world, and viticulture is moving towards more environmentally friendly production methods.

In this study, we assessed the impact of different organic soil amendments on selected soil physical–chemical and biochemical properties of the Sangiovese cultivar (*Vitis vinifera* L.). The results reported here offer new information regarding the impact of a new zeolite-based compost amendment on soil health in a perennial crop, such as vine. The research demonstrated that this new compost, especially at the lower concentration of zeolite 10 ZEO, could effectively be used to improve available water content, nutrient retention, carbon sequestration, and biological activity.

Comparing zeolite-based compost (ZEO treatments) with commercial compost (COM treatment), the higher mineralization rate in COM could lead to excess N in the soil. On the one hand, the increased N availability can stimulate plant growth, a positive feature for building vine structure; on the other hand, it can reduce the yield and quality of the grapes. Similarly, the higher total and lower available nutrients in ZEO-treated soils imply that phosphorus and potassium are released slowly over time, contributing to sustaining plant vigor.

In particular, 10 ZEO is the amendment that better combines an improvement in soil biochemical properties with gradual and constant nutrient availability, thus satisfying soil and plant needs without exceeding nutrient availability.

Based on these preliminary results, a positive long-term effect of ZEO treatments on improving vineyard soil quality and fertility should be expected. In addition, this management system, based on recycling a by-product of the wine supply chain, enables the closure of the production cycle of waste material in agricultural environments.

Author Contributions: Conceptualization, S.D.; methodology, E.P.; validation, S.D., E.P. and C.M.; formal analysis, S.D., F.V. and D.M.; investigation, S.D., D.M., E.C. and M.G.; resources, G.M., G.B.M. and M.G.; data curation, S.D., D.M., E.C., F.V., C.M. and M.G.; writing—original draft preparation, S.D. and E.P.; writing—review and editing S.D., C.M., F.V. and E.P.; visualization E.P.; supervision, S.D.; project administration, S.D., G.M. and G.B.M.; funding acquisition, G.M. All authors have read and agreed to the published version of the manuscript.

Funding: The study was carried out within the framework of the LIFE EU project LIFE ZeoWine LIFE17 ENV/IT/000427 "ZEOlite and WINEry waste as an innovative product for wine production". The LIFE ZeoWine Project is co-funded with the contribution of the European Commission, under the LIFE Programme—Environment & Resources efficiency—Grant Agreement n. LIFE17 ENV/IT/000427.

Data Availability Statement: The raw data supporting the conclusions of this article will be made available by the authors on request.

Acknowledgments: The authors gratefully acknowledge the technical assistance of Irene Rosellini and Manuele Scatena for sample collection, preparation, and analyses.

Conflicts of Interest: The authors declare no conflicts of interest. The funders had no role in the design of the study; in the collection, analyses, or interpretation of data; in the writing of the manuscript; or in the decision to publish the results.

References

1. Toreti, A.; Bavera, D.; Acosta Navarro, J.; Cammalleri, C.; de Jager, A.; Di Ciollo, C.; Hrast Essenfelder, A.; Maetens, W.; Magni, D.; Masante, D.; et al. *Drought in Europe August 2022*; Publications Office of the European Union: Luxembourg, 2022. [CrossRef]
2. Salomé, C.; Coll, P.; Lardo, E.; Metay, A.; Villenave, C.; Marsden, C.; Blanchart, E.; Hinsinger, P.; Le Cadre, E. The soil quality concept as a framework to assess management practices in vulnerable agroecosystems: A case study in Mediterranean vineyards. *Ecol. Indic.* **2016**, *61*, 456–465. [CrossRef]
3. Verheijen, F.G.A.; Jones, R.J.A.; Rickson, R.J.; Smith, C.J. Tolerable versus actual soil erosion rates in. *Europe Earth Sci. Rev.* **2009**, *94*, 23–38. [CrossRef]
4. Marinari, S.; Masciandaro, G.; Ceccanti, B.; Grego, S. Influence of organic and mineral fertilisers on soil biological and physical properties. *Bioresour. Technol.* **2000**, *72*, 9–17. [CrossRef]
5. Liu, X.; Liu, J.; Xing, B.; Herbert, S.J.; Meng, K.; Han, X.; Zhang, X. Effects of Long-Term Continuous Cropping, Tillage, and Fertilization on Soil Organic Carbon and Nitrogen of Black Soils in China. *Commun. Soil Sci. Plant Anal.* **2005**, *36*, 1229–1239. [CrossRef]
6. Biddoccu, M.; Zecca, O.; Audisio, C.; Godone, F.; Barmaz, A.; Cavallo, E. Assessment of long-term soil erosion in a mountain vineyard, Aosta Valley (NW Italy). *Land Degrad. Dev.* **2018**, *29*, 617–629. [CrossRef]
7. Kabasiita, J.K.; Opolot, E.; Malinga, G.M. Quality and Fertility Assessments of Municipal Solid Waste Compost Produced from Cleaner Development Mechanism Compost Projects: A Case Study from Uganda. *Agriculture* **2022**, *12*, 582. [CrossRef]
8. OIV State of the World Vine and Wine Sector 2021. 2022. Available online: https://www.oiv.int/public/medias/8778/eng-state-of-the-world-vine-and-wine-sector-april-2022-v6.pdf (accessed on 3 June 2022).
9. Chowdhary, P.; Gupta, A.; Gnansounou, E.; Pandey, A.; Chatuverdi, P. Current trends and possibilities for explotation of grape pomace as a potential source for value addition. *Environ. Pollut.* **2021**, *278*, 116796. [CrossRef]
10. Garcia-Perez, J.V.; Blasco, M.; Carcel, J.A.; Clemente, G.; Mulet, A. Drying Kinetics of Grape Stalk. In *Defect and Diffusion Forum*; Trans Tech Publications Ltd.: Baech, Switzerland, 2006; Volume 258–260, pp. 225–230.
11. Zhou, G.; Qiu, X.; Chen, L.; Zhang, C.; Ma, D.; Zhang, J. Succession of organics metabolic function of bacterial community in response to addition of earthworm casts and zeolite in maize straw composting. *Bioresour. Technol.* **2019**, *280*, 229–239. [CrossRef]
12. Gholamhoseini, M.; Aghaalikhani, M.; Dolatabadian, A.; Joghan, A.K.; Zakikhani, H. Decreasing N leaching and increasing canola forage yield in a sandy soil by application of natural zeolite. *Agron. J.* **2012**, *104*, 1467–1475. [CrossRef]
13. Soudejani, H.T.; Kazemian, H.; Inglezakis, V.J.; Zorpas, A.A. Application of zeolites in organic waste composting: A review. *Biocatal. Agric. Biotechnol.* **2019**, *22*, 101396. [CrossRef]
14. Manu, M.K.; Wang, C.; Li, D.; Varjani, S.; Wong, J.W.C. Impact of zeolite amendment on composting of food waste digestate. *J. Clean. Prod.* **2022**, *371*, 133408. [CrossRef]
15. Joghan, A.K.; Ghalavand, A.; Aghaalikhani, M.; Gholamhoseini, M.; Dolatabadian, A. Comparison among different integrated nutrition management for soil micro and macro elements after winter Wheat harvesting and yield. *Not. Sci. Biol.* **2010**, *2*, 107–111. [CrossRef]
16. Gholamhoseini, M.; Ghalavand, A.; Khodaei-Joghan, A.; Dolatabadian, A.; Zakikhani, H.; Farmanbar, E. Zeolite-amended cattle manure effects on sunflower yield, seed quality, water use efficiency and nutrient leaching. *Soil Tillage Res.* **2013**, *126*, 193–202. [CrossRef]
17. Mardini, B.; Shibusawa, S.; Kojima, Y.; Hosaka, S. Effect of natural zeolite (clinoptilolite) on ammonia emissions of leftover food-rice hulls composting at the initial stage of the thermophilic process. *J. Agric. Meteorol.* **2016**, *72*, 12–19. [CrossRef]
18. Campisi, T.; Abbondanzi, F.; Faccini, B.; Di Giuseppe, D.; Malferrari, D.; Coltorti, M.; Laurora, A.; Passaglia, E. Ammonium-charged zeolitite effects on crop growth and nutrient leaching: Greenhouse experiments on maize (*Zea mays*). *Catena* **2016**, *140*, 66–76. [CrossRef]
19. Doni, S.; Peruzzi, E.; Manzi, D.; Masini, C.M.; Mattii, G.B.; Macci, C.; Masciandaro, G. Co-composting winery waste and zeolite: A sustainable valorisation example. *J. Mater. Cycles Waste Manag.* **2023**. [CrossRef]
20. Soil Survey Staff. *Keys to Soil Taxonomy*, 12th ed.; Natural Resources Conservation Service, United States Department of Agriculture: Washington, DC, USA, 2014.

21. Olsen, S.R.; Sommers, L.E. Phosphorus. In *Methods of Soil Analysis: Part 2. Chemical and Microbiological Properties*, 2nd ed.; Page, A.L., Ed.; ASA and SSSA: Madison, WI, USA, 1982; pp. 403–430.
22. Helmke, P.A.; Sparks, D.L. Lithium, sodium, potassium, rubidium, and cesium. *Methods Soil Anal. Part 3 Chem. Methods* **1996**, *5*, 551–574.
23. Jakubus, M.; Spychasky, W. Evaluation of agricultural value of composts prepared from municipal biowastes in different conditions of composting process. *Agronomy* **2022**, *12*, 1438. [CrossRef]
24. Sumner, M.E.; Miller, W.P. Cation exchange capacity, and exchange coefficients. In *Methods of Soil Analysis. Part 2: Chemical Properties*, 3rd ed.; Sparks, D.L., Ed.; ASA, SSSA, CSSA: Madison, WI, USA, 1996.
25. Marx, M.C.; Wood, M.; Jarvis, S.C. A microplate fluorimetric assay for the study of enzyme diversity in soils. *Soil Biol. Biochem.* **2001**, *33*, 1633–1640. [CrossRef]
26. Vepsäläinen, M.; Kukkonen, S.; Vestberg, M.; Sirviö, H.; Niemi, R.M. Application of soil enzyme activity test kit in a field experiment. *Soil Biol. Biochem.* **2001**, *33*, 1665–1672. [CrossRef]
27. Richards, L.A. *Diagnosis and Improvement of Saline and Alkali Soil*; Handbook 60; US Salinity Lab. Staff, US Department of Agriculture, Agricultural Research Service: Washington DC, USA, 1954; 160p.
28. Ollat, N.; Peccoux, A.; Papura, D.; Esmenjaud, D.; Marguerit, E.; Tandonnet, J.P.; Bordenave, L.; Cookson, S.J.; Barrieu, F.; Rossdeutsch, L.; et al. Rootstocks as a component of adaptation to environment. In *Grapevine in a Changing Environment: A Molecular and Ecophysiological Perspective Eds Gers*; John Wiley & Sons: Hoboken, NJ, USA, 2016; pp. 68–108. [CrossRef]
29. Cramer, G.R.; Ergül, A.; Grimplet, J.; Tillett, R.L.; Tattersall, E.A.R.; Bohlman, M.C.; Vincent, D.; Sonderegger, J.; Evans, J.; Osborne, C.; et al. Water and salinity stress in grapevines: Early and late changes in transcript and metabolite profiles. *Funct. Integr. Genom.* **2007**, *7*, 111–134. [CrossRef]
30. Walker, R.R.; Blackmore, D.H.; Clingeleffer, P.R.; Correll, R.L. Rootstock effects on salt tolerance of irrigated field-grown grapevines (Vitis vinifera L cv Sultana): 1 yield vigour inter-relationships. *Aust. J. Grape Wine Res.* **2002**, *8*, 3–14. [CrossRef]
31. Chan, M.T.; Selvam, A.; Wong, J.W.C. Reducing nitrogen loss and salinity during "struvite" food waste composting by zeolite amendment. *Bioresour. Technol.* **2016**, *200*, 838–844. [CrossRef]
32. Yousefian, M.; Jafari, M.; Tavili, A.; Arzani, H.; Jafarian, Z. The effects of superabsorbent polymer on Atriplex lentiformis growth and soil characteristics under drought stress (case study: Desert research Station, Semnan, Iran). *J. Rangel Sci.* **2018**, *8*, 65–75.
33. Doni, S.; Gispert, M.; Peruzzi, E.; Macci, C.; Mattii, G.B.; Manzi, D.; Masini, C.M.; Masciandaro, G. Impact of natural zeolite on chemical and biochemical properties of vineyard soils. *Soil Use Manag.* **2020**, *37*, 832–842. [CrossRef]
34. Bastida, F.; Torres, I.F.; Hernandez, T.; García, C. The impacts of organic amendments: Do they confer stability against drought on the soil microbial community? *Soil Biol. Biochem.* **2017**, *113*, 173–183. [CrossRef]
35. Bastida, F.; García, C.; Fierer, N.; Eldridge, D.J.; Bowker, M.A.; Abades, S.; Alfaro, F.D.; Asefaw Berhe, A.; Cutler, N.A.; Gallardo, A.; et al. Global ecological predictors of the soil priming effect. *Nat. Commun.* **2019**, *10*, 3481. [CrossRef] [PubMed]
36. Guenet, B.; Camino-Serrano, M.; Ciais, P.; Tifafi, M.; Maignan, F.; Soong, J.L.; Janssens, I.A. Impact of priming on global soil carbon stocks. *Glob. Chang. Biol.* **2018**, *24*, 1873–1883. [CrossRef] [PubMed]
37. Georgiou, K.; Jackson, R.B.; Vindušková, O.; Abramoff, R.Z.; Ahlström, A.; Feng, W.; Harden, J.W.; Pellegrini, A.F.A.; Polley, H.W.; Soong, J.L.; et al. Global stocks and capacity of mineral-associated soil organic carbon. *Nat. Commun.* **2022**, *13*, 3797. [CrossRef] [PubMed]
38. Islam, M.R.; Singh, B.; Dijkstra, F.A. Stabilisation of soil organic matter: Interactions between clay and microbes. *Biogeochemistry* **2022**, *160*, 145–158. [CrossRef]
39. Kavvadias, V.; Ioannou, Z.; Vavoulidou, E.; Paschalidis, C. Short Term Effects of Chemical Fertilizer, Compost and Zeolite on Yield of Lettuce, Nutrient Composition and Soil Properties. *Agriculture* **2023**, *13*, 1022. [CrossRef]
40. Brombin, V.; Salani, G.M.; De Feudis, M.; Mistri, E.; Precisvalle, N.; Bianchini, G. Soil Organic Carbon Depletion in Managed Temperate Forests: Two Case Studies from the Apennine Chain in the Emilia-Romagna Region (Northern Italy). *Environments* **2023**, *10*, 156. [CrossRef]
41. Truc, M.T.; Yoshida, M. Effect of Zeolite on the Decomposition Resistance of Organic Matter in Tropical Soils under Global Warming. *Int. J. Innov. Sci. Res.* **2011**, *5*, 664–668.
42. Aminiyan, M.M.; Safari Sinegani, A.A.; Sheklabadi, M. The effect of zeolite and some plant residues on soil organic carbon changes in density and soluble fractions: Incubation study. *Eurasian J. Soil Sci.* **2016**, *5*, 74. [CrossRef]
43. Guimarães, D.V.; Gonzaga, M.I.; da Silva, T.O.; da Silva, T.L.; da Silva Dias, N.; Matias, M.I. Soil organic matter pools and carbon fractions in soil under different land uses. *Soil Tillage Res.* **2013**, *126*, 177–182. [CrossRef]
44. Beqiraj, E.; Gjoka, F.; Muller, F.; Baillif, P. Use of zeolitic material from Munella Region (Albania) as fertilizer in the sandy soils of Divjaka Region (Albania). *Carpth. J. Earth Environ. Sci.* **2008**, *3*, 33–47.
45. Allen, A.S.; Schlesinger, W.H. Nutrient limitations to soil microbial biomass and activity in loblolly pine forests. *Soil Biol. Biochem.* **2004**, *36*, 581–589. [CrossRef]
46. Conde, E.; Cardenas, M.; Ponce-Mendoza, A.; Dendooven, L. The impacts of inorganic nitrogen application on mineralization of 14C-labelled maize and glucose, and on priming ef- fect in saline alkaline soil. *Soil Biol. Biochem.* **2005**, *37*, 681–691. [CrossRef]
47. Baldi, E.; Polidori, G.; Germani, M.; Larocca, G.N.; Mazzon, M.; Allegro, G.; Pastore, C.; Quartieri, M.; Marzadori, C.; Filippetti, I.; et al. Fertilizer Potential of Organic-Based Soil Amendments on cv. Sangiovese (*V. vinifera* L.) Vines: Preliminary Results. *Agronomy* **2022**, *12*, 1604. [CrossRef]

48. Cataldo, E.; Salvi, L.; Paoli, F.; Fucile, M.; Masciandaro, G.; Manzi, D.; Masini, C.M.; Mattii, G.B. Application of zeolites in agriculture and other potential uses: A review. *Agronomy* **2021**, *11*, 1547. [CrossRef]
49. Sikora, J.L. Effect of compost-fertilizer blends on crop growth. In *The Science of Composting*; Bertoldi, M., Sequi, P., Lammers, B. Papi, T., Eds.; Blackie Academic & Professional Publications: Glasgow, UK, 1995; pp. 423–430.
50. Baerlocher, C.H.; Meier, W.M.; Olson, D.H. *Atlas of Zeolite Framework Types*, 5th ed.; Published on behalf of the Structure Commission of the International Zeolite Association by Elsevier: Amsterdam, The Netherlands, 2001; p. 299.
51. Hedström, A. Ion exchange of ammonium in zeolites: A literature review. *J. Environ. Eng.* **2001**, *127*, 673–681. [CrossRef]
52. Loganathan, P.; Vigneswaran, S.; Kandasamy, J., Bolan, N.S. Removal and re-recovery of phosphate from water using sorption *Crit. Rev. Environ. Sci. Technol.* **2014**, *44*, 847–907. [CrossRef]
53. Chen, K.; Zhao, K.; Zhang, H.; Sun, Q.; Wu, Z.; Zhou, Y.; Zhong, Y.; Ke, F. Phosphorus removal from aqueous solutions using a synthesized adsorbent prepared from mineralized refuse and sewage sludge. *Environ. Technol.* **2013**, *34*, 1489–1496. [CrossRef] [PubMed]
54. Meng, S.; Li, Y.; Zhang, T.; Chen, J.; Xu, P.; Song, C.; Fan, L.; Qiu, L. Influences of environmental factors on lanthanum/aluminum-modified zeolite adsorbent (La/Al-ZA) for phosphorus adsorption from wastewater. *Water Air Soil Pollut.* **2013**, *224*, 1556 [CrossRef]
55. Onyango, M.S.; Kuchar, D.; Kubota, M.; Matsuda, H. Adsorptive removal of phosphate ions from aqueous solution using synthetic zeolite. *Ind. Eng. Chem. Res.* **2007**, *46*, 894–900. [CrossRef]
56. Ozbahce, A.; Tari, A.F.; Gonulal, E.; Simsekli, N. Zeolite for enhancing yield and quality of potatoes cultivated under water-deficit conditions. *Potato Res.* **2018**, *61*, 247–259. [CrossRef]
57. Ramesh, K.; Biswas, A.K.; Somasundaram, J.; Rao, A.S. Nanoporous zeolites in farming: Current status and issues ahead. *Curr. Sci.* **2010**, *99*, 760–765.
58. Stylianou, M.A.; Inglezakis, V.J.; Loizidou, M.D. Effects of Zeolite Addition on Soil Chemistry-Open Field Experiments. In Proceedings of the Protection and Restoration of The Environment VII, Mykonos, Greece, 28 June–1 July 2004; pp. 1–7.
59. Hazrati, S.; Tahmasebi-Sarvestani, Z.; Modarres-sanavy, S.A.M.; Mokhtassi-Bidgoli, A.; Mohammadi, H.; Nicola, S. Effects of zeolite water stress on growth: Yield chemical compositions of *Aloe vera*, L. *Agric. Water Manag.* **2017**, *181*, 66–72. [CrossRef]
60. Ippolito, J.A.; Tarkalson, D.D.; Lehrsch, G.A. Zeolite Soil Application Method Affects Inorganic Nitrogen, Moisture, and Corn Growth. *Soil Sci.* **2011**, *176*, 136–142. [CrossRef]
61. Nakhli, S.A.A.; Delkash, M.; Bakhshayesh, B.E.; Kazemian, H. Application of zeolites for sustainable agriculture: A review on water and nutrient retention. *Water Air Soil Pollut.* **2017**, *228*, 464. [CrossRef]
62. Six, J.; Feller, C.; Denef, K.; Ogle, S.M.; de Morales, J.C.; Albrecht, A. Soil organic matter, biota and aggregation in temperate and tropical soils: Effects of no-tillage. *Agronomie* **2002**, *22*, 755–775. [CrossRef]
63. Gil-Sotres, F.; Trasar-Cepeda, C.; Leirós, M.C.; Seoane, S. Different approaches to evaluating soil quality using biochemical properties. *Soil Biol. Biochem.* **2005**, *37*, 877–887. [CrossRef]
64. Wittmann, C.; Kähkönen, M.A.; Ilvesniemi, H.; Kurola, J.; Salkinoja-Salonen, M.S. Areal activities and stratification of hydrolytic enzymes involved in the biochemical cycles of carbon, nitrogen, sulphur and phosphorus in podsolized boreal forest soils. *Soil Boil. Biochem.* **2004**, *36*, 425–433. [CrossRef]
65. Yousefi, R.; Amooei, A.A.; Amel Sakhi, M.; Karimi, A. Experimental study on influence of using urease enzyme on stabilized sandy soil's engineering property by zeolite and sawdust. *Int. J. Marit. Technol.* **2021**, *15*, 17–27.
66. Peng, X.Q.; Wang, W. Stoichiometry of soil extracellular enzyme activity along a climatic transect in temperate grasslands of northern China. *Soil Biol. Biochem.* **2016**, *98*, 74–84. [CrossRef]
67. Liu, E.; Zhou, J.; Yang, X.; Jin, T.; Zhao, B.; Li, L.; Wen, Y.; Soldatova, E.; Zamanian, K.; Gopalakrishnan, S.; et al. Long-term organic fertilizer-induced carbonate neoformation increases carbon sequestration in soil. *Environ. Chem. Lett.* **2023**, *21*, 663–671. [CrossRef]
68. Zhou, J.; Wen, Y.; Shi, L.; Marshall, M.R.; Kuzyakov, Y.; Blagodatskaya, E.; Zang, H. Strong priming of soil organic matter induced by frequent input of labile carbon. *Soil Biol. Biochem.* **2021**, *152*, 108069. [CrossRef]
69. Eivazi, F.; Tabatabai, M. Factors affecting glucosidase and galactosidase in soils. *Soil Biol. Biochem.* **1990**, *22*, 891–897. [CrossRef]
70. Condron, L.M.; Turner, B.L.; Cade-Menun, B.J.; Sims, J.; Sharpley, A. Chemistry and dynamics of soil organic phosphorus. In *Phosphorus: Agriculture and the Environment*; John Wiley & Sons, Inc.: Hoboken, NJ, USA, 2005; pp. 87–121.
71. Peng, Y.P.; He, W.X.; Wang, Z.Q.; Zhang, J.; Yan, H.R.; Gao, Y.J.; Cao, W.D. Soil chemical properties and enzyme activities in long-term green manure plot in Loess Plateau. *J. Northwest AF Univ. (Nat. Sci. Ed.)* **2015**, *43*, 131–138. (In Chinese)
72. Delgado, M.; Rodríguez, C.; Martín, J.V.; Miralles de Imperial, R.; Alonso, F. Environmental assay on the effect of poultry manure application on soil organisms in agroecosystem. *Sci. Total Environ.* **2012**, *416*, 532–535. [CrossRef]
73. Chen, H.; Luo, P.; Wen, L.; Yang, L.Q.; Wang, K.L.; Li, D.J. Determinants of soil extracellular enzyme activity in a karst region southwest China. *Eur. J. Soil Biol.* **2017**, *80*, 69–76. [CrossRef]

environments

MDPI

Article

The Application of Cyanobacteria as a Biofertilizer for Okra (*Abelmoschus esculentus*) Production with a Focus on Environmental and Ecological Sustainability

Saoli Chanda [1,2], Sanku Dattamudi [1,2,*], Krishnaswamy Jayachandran [1], Leonard J. Scinto [3] and Mahadev Bhat [1]

1 Agroecology Program, Department of Earth and Environment, Florida International University, Miami, FL 33199, USA; schanda@fiu.edu (S.C.); jayachan@fiu.edu (K.J.); bhatm@fiu.edu (M.B.)
2 Department of Agriculture, Agribusiness, & Environmental Sciences, Texas A&M University-Kingsville, Kingsville, TX 78363, USA
3 Department of Earth and Environment, Florida International University, Miami, FL 33199, USA; scintol@fiu.edu
* Correspondence: sanku.dattamudi@tamuk.edu

Abstract: Cyanobacteria, an important addition to biofertilizers, are gaining popularity for their multifaceted benefits in sustainable agriculture and ecosystem restoration. However, harmful algal blooms (HABs) in freshwater, predominantly caused by cyanobacteria, prevent sunlight penetration into the water and develop hypoxic and anoxic conditions. We collected cyanobacteria slurry from Lake Jesup (Central Florida, USA), repurposed it as a biofertilizer, and incorporated it in a typical South Florida calcite soil for high-value okra (*Abelmoschus esculentus*; var: Clemson spineless) production. Experiments were conducted at the Organic Garden Shade House and Greenhouse located inside the main campus of the Florida International University (FIU), FL, USA. A two-year experiment with four different treatments was conducted, namely, (a) control (C; no fertilizer applied), (b) total synthetic (TS), (c) total biofertilizer (TB; only cyanobacteria biofertilizer was applied), and (d) half and half (HH; 50% biofertilizer + 50% synthetic fertilizer), which were arranged in a randomized complete block design (RCBD) with six replications for each treatment. Our results indicate that TB and TS produced about 29 to 33% higher SPAD (soil plant analytical development) readings than the control. The absence of interveinal chlorosis (yellowing of leaves) in the TB and HH treatments suggests that the cyanobacteria-based biofertilizer had a role in supplying one of the critical micronutrients, iron (Fe). Analysis of the biofertilizer indicated 2000 ppm Fe content, which directly supports our observation. Similarly, average plant height (61 cm), yield (130 gm per pot), and crop biomass (67 gm) productions were significantly higher in TB than in the control. Overall, this study documents the potential of cyanobacteria biofertilizers as a viable option compared to synthetic fertilizers for sustainable crop production and soil health improvement.

Keywords: cyanobacteria; biofertilizer; okra; organic nutrient source; Florida; sustainable agriculture

1. Introduction

Biofertilizers are organic substances containing abundant bio-stimulants and microorganisms. Biofertilizers include algae, plant growth-promoting rhizobacteria (PGPR), and other mineralizing microorganisms, all of which can potentially contribute to mobilizing nutrients in the soil, thereby enhancing soil and crop productivity. Additionally, biofertilizers are considered organic substitutes for synthetic chemical fertilizers to promote organic farming practices. Cyanobacteria (prokaryotic microorganisms), a group of photosynthesizing bacteria, are gaining acceptance as organic biofertilizers due to their multifaceted benefits to soil health and crop production. Unlike other symbiotic microorganisms, cyanobacteria do not depend on a host for their growth and development and have a nitrogen-fixing capability of 25 to 60 kg N ha^{-1} in the soil [1]. Some of the benefits of cyanobacteria biofertilizers are (a) the release of extracellular polymeric substances (polysaccharides) to improve soil aggregate stability, the (b) release of plant growth-promoting hormones [2], (c) the increase in leaf chlorophyll content and antioxidant enzyme activity, and (d) maintaining a symbiotic relationship with other microorganisms and enhancing microbial biomass C in soil [3]. The biofilm layer (from biological soil crusts; BSCs) formed from the extracellular polysaccharides of cyanobacteria serves as a habitat for other organisms in the soil. This synergistic relationship often improves soil biodiversity and ecosystem resiliency. Cyanobacteria are also capable of sequestering carbon in their cells (like C_4 plants) [4].

However, the presence of cyanobacteria in freshwater ecosystems causes a major threat by creating harmful algal blooms (HABs). HABs prevent sunlight penetration in water and create eutrophic conditions. Once the algal blooms complete their life cycle and decompose, secondary microbes decompose the biomass, resulting in oxygen depletion. In addition to creating hypoxic conditions in waterbodies, cyanobacteria produce cyanotoxins, which are harmful to humans, animals, and aquatic organisms. Specifically, favorable weather conditions (warm temperature and high relative humidity) in Florida have led to high HABs in major lakes, generally covering 60–80% of the lake area during peak algal bloom season (May to mid-September). For the last couple of decades, the occurrence of harmful algal blooms has been a recurrent problem in Florida, causing environmental and ecological degradation and economic loss. One of the remedial measures is the unwanted cyanobacteria slurry collected from HABs disposed of as a landfill material. In this study, we repurposed the harvested cyanobacteria slurry to produce biofertilizers for growing high-value organic vegetables. Specifically, we used cyanobacteria biofertilizers for okra (specialty crop) cultivation. Okra is one of the most important vegetable crops grown in Florida, contributing a significant revenue to the state economy. Okra is grown approximately 1000 to 1500 acres annually in Miami-Dade County, South Florida [5]. Okra is also a short-duration crop (generally takes 65 to 70 days from planting to harvesting) and prefers hot weather, which makes it easier to grow okra during the summer months in South Florida.

The application of cyanobacteria as a biofertilizer in row crops, such as rice, wheat, soybean, cotton, and maize, has been well documented [6,7]. However, most of these studies have been conducted either directly in the field (such as a paddy rice field) or used cyanobacteria inoculants for their experiments. Studies focusing on the application of cyanobacteria as a biofertilizer (after drying and processing) for producing crops and maintaining soil quality are still underexplored. Furthermore, there is a lack of understanding of the beneficial effects of cyanobacteria as a biofertilizer on crop production and soil health in the United States. Typically, chicken manure, turkey manure, pig/cow manure, vermicompost, and other composting materials are used for organic farming in the USA. To the best of our knowledge, no experiments conducted in the USA have explored the effect of naturally occurring cyanobacteria as a biofertilizer for crop production and soil health improvement. Therefore, our research objective was to evaluate the performance of cyanobacteria biofertilizers in the geographical and climatic conditions (sub-tropical) of South Florida. We also expect that our study will improve the basic knowledge about the numerous benefits that this biofertilizer can offer in an agricultural production system. The specific objectives of this study were 1) to evaluate the effects of cyanobacteria biofertilizers on plant growth and physiological parameters of okra and 2) to assess the efficacy of cyanobacteria biofertilizers for improving soil properties compared to synthetic fertilizers.

2. Materials and Methods

2.1. Cyanobacteria Collection and Biofertilizer Processing

Cyanobacteria slurry materials were collected from Lake Jesup (28°43′ N, 81°13′ W) located in Central Florida, USA. The slurry was collected from different locations of the Lake. The cyanobacterial slurry was collected in five-gallon buckets for easy transportation and brought back to the University campus for further processing. Large plant roots and other visible unwanted things were discarded carefully from the collected materials. The collected slurry was air dried (after spreading on large plastic trays) in the greenhouse for 3 to 4 weeks to obtain dried cyanobacteria materials (Figure 1). The slurry on the plastic trays was stirred up occasionally to expedite the drying process. Dried cyanobacteria were then powdered using a mechanical grinder and stored in colored plastic bags for future applications. Storing cyanobacteria biofertilizers in colored plastic bags were reported to increase shelf life of the biofertilizer [8]. Cyanobacteria can also be used in the soil as semi-solid wet materials; however, it is difficult to measure the nutrient concentrations in wet cyanobacteria, and the wet materials need to be applied immediately after collection. Dried and powdered cyanobacteria materials were mixed with dried soil before filling up the pots for a uniform distribution of the biofertilizer. A recent study reported that dry cyanobacteria material works more effectively as fertilizer than wet materials for crop production [9]. The concentrated slurry had about 72 to 75% water (roughly 12 to 15% dry matter content). The dry matter content of the slurry was processed (after drying) to produce the biofertilizer.

2.2. Experimental Design and Sample Collection

This two-year experiment was conducted in the greenhouse (year one) and in the shade house (year two) at Florida International University (FIU), Florida, USA. Soils (Biscayne marly silt loam) used for this experiment were collected from a farmer's field located in Homestead, FL. The physicochemical properties of the background soil are presented in Table 1. Okra (*Abelmoschus esculentus*; var: Clemson spineless) was grown in three-gallon (11.4 L) plastic nursery pots spaced 2.5 ft (0.91 m) apart from each other (Figure 1D). Four experimental treatments, namely, the (a) control (C; no fertilizer applied), (b) total synthetic (TS; synthetic urea 46-0-0 and sulfate of potash 0-0-51 were applied), (c) total biofertilizer (TB; only cyanobacteria biofertilizer was applied), and (d) half and half (HH; 50% biofertilizer + 50% synthetic fertilizer were applied), were assigned in a randomized complete block design (RCBD) with six replications for each treatment for both year one and two experiments. Recommended doses of N and K for okra production are 135 kg N ha^{-1} and 112 kg K_2O ha^{-1}. The amount of biofertilizer needed for each treatment (TB and HH) was calculated based on the nutrient contents of the dried cyanobacteria biofertilizer (Table 2). Composite soil samples (0 to 15 cm depth) collected from the field for these two experiments were analyzed for physicochemical properties. It can be noted that the collected soil was an actual representation of typical South Florida organic soils, which have alkaline limestone parent material (high $CaCO_3$). Large roots and stones were removed from the soil through sieving before filling the pots.

Figure 1. Pictures of okra growth in greenhouse settings. (**A**) Cyanobacteria slurries collected in five-gallon buckets, (**B**) drying of cyanobacteria materials, (**C**) one-week-old okra seedlings, (**D**) greenhouse settings of the experiment (including all treatments), (**E**) okra flowers, (**F**) okra flower and fruits together, and (**G**) early nutrient deficiency in control plants.

Table 1. Physicochemical properties of the soil used for the experiments.

Parameters	Unit	Value
pH		8.01
Total C (TC)	%	6.07 ± 0.42
Total N (TN)	%	0.34 ± 0.06
Total P (TP) *	ppm	103 ± 27
Potassium (K)	ppm	175 ± 42
Calcium (Ca)	ppm	13,604 ± 2722
Magnesium (Mg)	ppm	255 ± 29
Sulphur (S)	ppm	79.20 ± 8.24
Zinc (Zn)	ppm	24.48 ± 5.69
Copper (Cu)	ppm	32.80 ± 4.91
Sodium (Na)	ppm	51.73 ± 11.73

* Analyzed using the Solórzano and Sharp (1980) method and the EPA method 365.1. Analyzed through Mehlich 3 extraction.

Table 2. Physicochemical properties of the dried and powdered cyanobacteria biofertilizer.

Parameters	Unit	Value
Total C (TC)	%	19.56 ± 0.71
Total N (TN)	%	1.79 ± 0.07
Total P (TP)	%	0.02 ± 0.00
Total S (TS)	%	0.13 ± 0.01
Potassium (K)	%	0.06 ± 0.01
Calcium (Ca)	%	6.12 ± 0.70
Magnesium (Mg)	%	0.12 ± 0.01
Iron (Fe)	ppm	2005 ± 160
Manganese (Mn)	ppm	132.56 ± 13.35
Zinc (Zn)	ppm	53.34 ± 0.69
Copper (Cu)	ppm	29.51 ± 4.80
Boron (B)	ppm	95.37 ± 4.21
Molybdenum (Mo)	ppm	1.82 ± 1.01
Nickel (Ni)	ppm	4.34 ± 1.78

Three okra seeds were planted for each pot, and thinning was performed at the four-leaf stage to keep the healthiest plant in the pot. Soils from each pot were also collected at harvesting and analyzed for physicochemical properties.

2.3. Plant Height, Stem Diameter, Leaf Chlorophyll Content, Crop Biomass, and Okra Fruit Yield

Plant height (cm), stem diameter (cm), and leaf chlorophyll content of the okra plant from each pot were recorded at four, five, six, and seven weeks after planting and at harvesting. The experiment was terminated at 67 days after sowing (DAS), and fruit yield was obtained by weighing the harvested okra from each pot. Plant height and stem diameter (five readings from each plant) were measured using meter sticks and slide calipers, respectively. The average leaf chlorophyll content of the developed upper leaves was recorded using the Soil-Plant Analyses Development (SPAD) 502 Plus Chlorophyll Meter (Konica Minolta, Inc, Ramsey, NJ, USA). Fresh shoot and root samples were collected during harvesting (67 DAS). Shoot samples (stem and leaves) were taken from the first node up. Roots were carefully washed to remove soil particles. Both shoot and root samples were dried at 60 °C for 72 h to obtain dry aboveground and belowground biomasses.

2.4. Laboratory Analyses

Soil pH was measured (1:1 dried ground soil to water ratio) using a seven excellence pH meter (Mettler Toledo, Columbus, OH, USA). Analysis of total nitrogen and total carbon content of the collected dried ground cyanobacteria, soil, and plant samples were performed using a CN analyzer (LECO company, St. Joseph, MI, USA) following the Dumas dry combustion method. The dried ground cyanobacteria and plant samples were digested using nitric acid–hydrogen peroxide to determine micro- and macronutrient contents. Later, the digested and filtered samples were analyzed using Perkin Elmer Optima 8300 ICP-OES (Waltham, MA, USA). Dried ground soil samples were extracted using the Mehlich 3 extraction procedure. The filtered samples were analyzed using Perkin Elmer Optima 8300 ICP-OES.

2.5. Statistical Analyses

All data in this study are presented as mean and standard deviation. Statistical analyses were performed using SAS 9.4 (SAS Institute NC, Cary, NC, USA). The treatment effects on different plant growth parameters, soils, and plant physicochemical properties were evaluated using a one-way analysis of variance (ANOVA). A mean comparison was performed using Tukey–Kramer and Fisher's LSD post hoc tests at $p < 0.10$ and $p < 0.05$. Differences at mean values $p < 0.05$ were considered statistically significant, while differences at mean values $p < 0.05$ were discussed.

3. Results and Discussion

3.1. Physicochemical Properties of the Biofertilizer

The N content in the cyanobacteria biofertilizer ranged from 1.56 to 1.94% (Table 2), which appears to be higher compared to the N content reported in a couple of published articles [10,11]. However, most of those previous studies produced cyanobacteria under controlled conditions (tunnel house, small microcosm) or cultured in a rice paddy field. We collected the free-flowing cyanobacteria slurries (naturally grown) from Lake Jesup in Central Florida. The lake often receives high amounts of nutrients as runoff from the adjacent agricultural areas [12], which resulted in higher nutrient contents in our cyanobacteria dry mix. Cyanobacteria adsorb or fix atmospheric N and store it in their vacuoles, but the magnitude of N fixation commonly depends on the cyanobacteria species. For instance, *Anabaena* and *Nostoc* cyanobacteria species used for biofertilizer preparation (when inoculated in a controlled environment) have higher N content than other species [1,13]. We collected free-flowing cyanobacteria from the lake where *Anabaena, Nostoc,* and *Clostridium* species were commonly present. It was difficult to segregate those species from the slurry mixture; therefore, it was expected that our samples had a mixture of different species of cyanobacteria.

The C:N ratio of our biofertilizer was in the range of 12:1 to 14:1, which is lower than organic wastes, like straw 100:1, sawdust 200:1, leaves, and plant stalks. Therefore, it takes months to process the organic wastes, such as straw and sawdust (to lower the C:N ratio), before application in the field. So, it was a major advantage of our newly developed biofertilizer where a lower C:N ratio can potentially influence the net N mineralization in the soil and thus, the biofertilizer was ready to be applied immediately in the soil as a source of plant nutrients.

Fe (>2000 ppm) and Mg (>1200 ppm) were in higher concentrations in the biofertilizer produced in this study (Table 2). South Florida soils are mostly porous sandy loam, with alkaline soil pH and low organic matter content. Thus, growers in South Florida (specifically organic growers) often face production issues due to Fe deficiency in the soil (personal communication with the growers), resulting from calcareous parent materials. Therefore, it is expected that the cyanobacteria biofertilizer developed in this study will be able to reduce the Fe deficiency problem in South Florida soils. More about the effects of Fe and Mg on okra production at different treatments are discussed later in this manuscript (Section 3.2(a)).

3.2. Plant Physiological Parameters

(a) Leaf chlorophyll content—SPAD values

Chlorophyll content in plant leaves is an indicator of leaf N content, chloroplast development, and overall plant health. The soil–plant analytical development (SPAD) meter is a simple, rapid, and non-destructive instrument used to analyze leaf chlorophyll contents of plants. SPAD generates three-digit readings (unitless), where higher SPAD values indicate high leaf N content and a healthy plant. We started collecting SPAD data four weeks after planting the crop when the leaves were old enough to produce SPAD readings. SPAD readings were recorded every week afterwards (until week 7) and at harvesting to obtain the leaf chlorophyll content of okra (Figure 2). In this study, SPAD values were highly variable throughout the experiment; however, prominent differences among treatments were observed at five and six weeks after planting the crop. Average SPAD values of TS and TB pots were 1.29 and 1.33 times higher, respectively (significant at $p < 0.05$), than control pots, indicating that plants in TS and TB pots had higher plant vigor and leaf chlorophyll content than pots that received no fertilizer treatment. A study conducted by Romanowska-Duda et al. [14] documented similar observations and found that cyanobacteria biofertilization increased leaf chlorophyll content, photosynthetic productivity, and crucial plant growth enzymes.

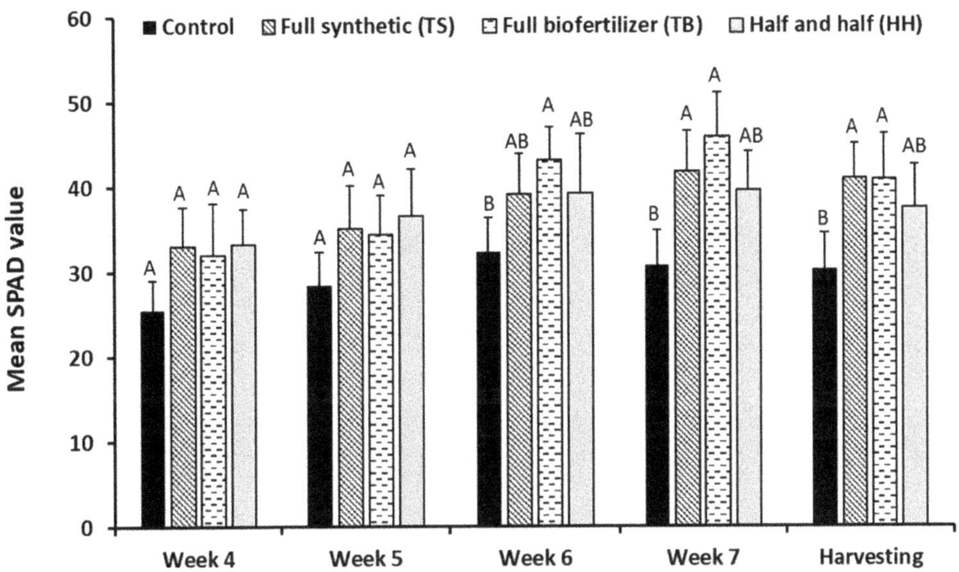

Figure 2. Soil–plant analytical development (SPAD) values of okra at different growth stages. Similar uppercase letters indicate no significant difference at $p < 0.10$.

SPAD values of control pots (average 29.44 for all five data points) started declining after week 6 (32.31), with a parallel observation of leaf yellowing in some pots Common leaf yellowing (chlorosis) of plants is an indication of nutrient (N, S, Fe, Mn) deficiency, depending on the location of the deficiency symptoms (older vs. younger leaves) and type of the chlorosis (full chlorosis or interveinal). However, a prominent pale-yellow leaf color (interveinal chlorosis) and stunted plant growth were observed 50 days after planting (mostly in younger leaves), indicating the possibility of Fe deficiency in those plants. Visual observation of the chlorosis is displayed in Figure 3. A sign of the yellowing of leaves was also observed in TS plants, but the number of leaves affected was fewer than in control plants. South Florida soils are deficient in Fe (due to alkaline soil pH), and deficiency symptoms are common among plants that received no fertilizer treatments (control pots). Comparing the treatments, no yellowing of leaves was observed in TB and HH pots, resulting from high Fe content (more than 2000 ppm) in the biofertilizer prepared from cyanobacteria (Table 2). The cyanobacteria biofertilizer was reported to provide a significant amount of Fe in the soil along with other nutrients [9]. Additionally, Mg, which is a main component of chlorophyll, was in abundant quantity (1200 ppm) in the cyanobacteria biofertilizer prepared for this experiment. To support our observation, we analyzed the plant samples from each pot after harvesting., Plant analysis data indicate that Fe content was significantly higher in TB (258 ppm) and HH (220 ppm) than in the control (152 ppm), further confirming the fact that chlorosis in the control plants was Fe deficiency symptoms.

Figure 3. Visual representation of possible Fe deficiency (interveinal chlorosis) in control plants from the shade house study. The top picture shows the difference between leaves from full cyanobacteria pots (green color) and control pots (light green color with pale yellow spots).

Average SPAD values for TS pots at weeks four and five were 2 to 4% higher than TB; however, starting from week 6, we observed higher SPAD values in TB pots than in TS pots. A common explanation is the synthetic fertilizers, specifically urea (producing ammonia after hydrolysis), which tend to be easily volatized from the soil profile [15], and thus, less amount of soil N was available for plants at five weeks after planting. A lower amount of N in the leaf produced lower SPAD values for TS pots.

The average SPAD value combining all treatments (excluding the control) at harvesting was 7% lower than the value at week seven, confirming that leaf senescence close to harvest would have resulted in a lower SPAD value. It should be noted that the SPAD value depends on various factors, including leaf age, environmental conditions, and other factors [16]. It is noted that lower SPAD values close to the crop harvesting stage were observed in previous studies [17,18].

(b) Plant height, stem diameter, and crop biomass production and okra yield

Average plant height, stem diameter, and crop biomass (shoot and root) are summarized in Table 3. Plant heights of TS (62.8 cm) and TB (61.3 cm) were similar to each other (not significantly different); however, both TS and TB had significantly higher plant heights than the control treatment. Plant height is an active indicator of vegetative growth or the vigor of the plant. The difference in plant height of okra under various treatments during weeks five and seven is easily visible in Figure 4.

Table 3. Physiological parameters (plant height, stem diameter, aboveground and belowground biomasses) and okra yield under different treatments. Experiments were conducted at the FIU Organic Garden Shade House and the FIU greenhouse. Data presented are an average of year one and year two experiments.

Treatments	Plant Height *	Stem Diameter *	Shoot Dry Weight¶	Root Dry Weight *	Shoot/Root	Yield¶
	cm	cm	g			g/pot
Control	45.3 ± 3.68 b	0.78 ± 0.08 a	26.71 ± 4.61 b	5.61 ± 0.88 b	4.69	73.38 ± 5.27 b
TS	62.8 ± 4.81 a	1.08 ± 0.16 a	47.92 ± 6.99 ab	8.37 ± 1.95 ab	6.14	120.98 ± 9.42 a
TB	61.3 ± 5.16 a	1.11 ± 0.13 a	55.01 ± 7.62 a	11.58 ± 2.16 a	3.78	130.34 ± 8.78 a
HH	58.3 ± 3.68 ab	1.05 ± 0.13 a	37.07 ± 3.92 ab	7.59 ± 0.78 ab	4.21	110.18 ± 12.36 ab

TS = only (100%) synthetic fertilizer; TB = only (100%) cyanobacteria biofertilizer; HH = half (50%) synthetic fertilizer and half (50%) cyanobacteria biofertilizer. * Similar lowercase letters indicate no significant difference at $p < 0.10$; ¶ similar lowercase letters indicate no significant difference at $p < 0.05$.

Stem diameter, which is also an indicator of plant water status [19], was highest in okra grown with TB (average 1.11 cm) followed by TS, HH, and the control. It is often described that a higher stem diameter indicates better plant growth [20] and can result in high crop biomass production. Consequently, a similar trend was observed for plant biomass production, where total biomass production (shoot and root combined) in TB and TS was 1.60 to 1.80 times higher than the control. An interesting observation was that the higher root biomass in TB pots (average 11.6 gm) was higher than in other treatments. The shoot to root ratio was the lowest in TB pots (3.78), indicating that the cyanobacteria biofertilizer was able to improve root structure more than shoot biomass production in this study. Biofertilizer pot (TB) soils at harvesting had numerically higher (196 ppm) total P content than synthetic pots (TS; 117 ppm), which was another possible reason for better root growth in plants that received biofertilizer treatments. A research study documented that cyanobacteria are capable of increasing root length, root volume, and quality for vegetable crops [21], which is a major advantage of those crops for efficient nutrient uptake from the soil.

Yield analysis of okra plants from four treatments is presented in Table 3. The average yields (gm per pot) of TB (130) and TS (121) were significantly higher ($p < 0.001$) than the control. Biofertilizers act as natural sources of plant nutrients and promote overall plant health and productivity. An interesting observation during these two experiments was quicker physiological developments (specifically plant height and stem diameter) of the plants that received biofertilizer treatments (both in TB and HH) than other treatments. Specifically, flowering and fruit settings in biofertilizer-treated pots were observed at least 7 to 10 days earlier than other treatments. We did not measure plant growth regulators and hormones in the biofertilizer soils, but previous studies reported that cyanobacteria exudates are capable of adding bioactive compounds and growth regulators in the soil [2, 22], which improve plant physiological parameters.

Figure 4. Visual representation of okra plants (shade house study) received different treatments (**A**) at week 4 and (**B**) at week 6 after planting. Treatments in this picture are as follows (from left to right): full synthetic fertilizer (TS), full cyanobacteria biofertilizer (TB), HH (half (50%) synthetic fertilizer and half (50%) cyanobacteria biofertilizer), and control. Pictures are from the shade house study.

3.3. *Soil Properties at Harvesting*

Soils collected after harvesting were analyzed for physicochemical properties (Table 4). A major observation was lower soil pH in biofertilizer-treated pots (both in TB and HH) than in other pots. The decomposition of cyanobacteria often releases organic acids in the soil during decomposition [2], which explains the slightly acidic pH in the TB soils at harvesting. Since South Florida soils are alkaline (with high $CaCO_3$ contents), cyanobacteria biofertilizer decomposition will help reduce the soil pH and solubilize the nutrients. Even for a shorter study (for about 70 days), the amount of C (10.77%) and N (0.53%) contents in soils at TB pots at harvesting were significantly higher than the control (C, 5.46% and N, 0.30%). Mg, which is a major component of chlorophyll, was also significantly higher in TB pots than other treatments. Phosphorus contents (ppm) in the soil at harvesting from TB (196) and HH (141) pots were significantly higher ($p < 0.05$) than control pots possibly because of the ability of the biofertilizer to enhance P bioavailability in the soil through solubilization. Overall, most of the nutrients were higher in TB soils at harvesting compared to the control, an indication of improved soil nutrient status in cyanobacteria-applied pots. In a study conducted in Nigeria, Agwa et al. [23] found that cyanobacteria inoculants significantly increased soil nutrient status for okra production.

Table 4. Okra leaves' nutrient content at harvest. Data presented are an average of year one and year two experiments.

Treatments	C	N	P	K	Ca	Mg	S	Fe	Mn	B
	%							ppm		
Control	27 ± 6 b	1.9 ± 0.4 b	0.18 ± 0.04 b	1.3 ± 0.41 a	2.9 ± 0.7 b	0.62 ± 0.18 a	0.32 ± 0.14 a	152 ± 27 c	17 ± 4 b	48 ± 11 a
TS	36 ± 11 ab	2.9 ± 0.6 ab	0.26 ± 0.07 ab	2.0 ± 0.62 a	4.1 ± 1.1 ab	0.85 ± 0.29 a	0.33 ± 0.14 a	173 ± 42 bc	37 ± 8 a	40 ± 9 a
TB	44 ± 7 a	3.7 ± 0.5 a	0.32 ± 0.07 a	2.3 ± 0.59 a	4.8 ± 1.0 a	0.95 ± 0.21 a	0.35 ± 0.17 a	258 ± 31 a	28 ± 11 ab	42 ± 13 a
HH	38 ± 10 ab	3.1 ± 0.6 a	0.22 ± 0.07 ab	1.7 ± 0.24 a	4.5 ± 0.9 ab	0.88 ± 0.27 a	0.30 ± 0.16 a	220 ± 29 ab	33 ± 7 a	51 ± 7 a

TS = only (100%) synthetic fertilizer; TB = only (100%) cyanobacteria biofertilizer; HH = half (50%) synthetic fertilizer and half (50%) cyanobacteria biofertilizer; C, N, P, K, Ca, Mg, and S are presented as %, and Fe, Mn, and B are presented as ppm. Similar lowercase letters indicate no significant difference at $p < 0.10$.

3.4. Environmental Implications

A major component of this research study was to foster environmental and ecological sustainability. We removed free-flowing cyanobacteria from the Lake, which is otherwise a great threat to the aquatic life due to eutrophication (reducing the dissolved oxygen content of the water). Additionally, a cyanobacteria slurry in the lake causes a nuisance (foul smell) to the adjacent neighborhood. Repurposing cyanobacteria as a biofertilizer in field-scale agricultural settings presents an opportunity to explore their potential benefits for crop production and soil quality improvement in diverse agroecosystems. The balanced array of plant nutrients, including both macro- and micronutrients, provided by the cyanobacteria biofertilizer (Table 2) offers an alternative to synthetic fertilizers for crop production in South Florida and beyond. As a result, it is expected to reduce nutrient runoff to the surface water systems and improve overall environmental quality. As mentioned earlier, South Florida soils are predominantly porous sandy loam, alkaline, and low in organic matter (OM) content. Our biofertilizer contributes organic carbon to the soil (OC content in cyanobacteria soil was two times higher than control soil; Table 5), which is expected to improve soil quality and aggregate stability. Better soil structure not only increases nutrient retention and availability to plants but also reduces nutrient leaching into freshwater systems, such as Lake Jesup and surrounding areas, further supporting environmental and ecological sustainability efforts.

Table 5. Soil nutrient content after okra harvesting. Data presented are an average of year one and year two experiments.

Treatments	pH	C	N	P	K	Ca	Mg	S	Na
		%		ppm					
Control	7.69	5.46 ± 0.49 b	0.296 ± 0.05 c	86 ± 19 b	154 ± 28 b	12,064 ± 1223 b	242 ± 37 b	37 ± 6 b	47 ± 15 a
TS	7.84	8.44 ± 1.27 a	0.438 ± 0.07 ab	117 ± 39 ab	388 ± 111 a	13,289 ± 1361 b	354 ± 38 a	102 ± 18 a	67 ± 14 a
TB	6.63	10.77 ± 1.53 a	0.533 ± 0.06 a	196 ± 41 a	301 ± 120 ab	20,850 ± 2209 a	384 ± 29 a	89 ± 11 a	82 ± 24 a
HH	7.89	9.15 ± 1.11 a	0.407 ± 0.05 b	141 ± 22 a	256 ± 62 a	13,927 ± 1621 b	325 ± 41 a	116 ± 22 a	74 ± 22 a

TS = only (100%) synthetic fertilizer; TB = only (100%) cyanobacteria biofertilizer; HH = half (50%) synthetic fertilizer and half (50%) cyanobacteria biofertilizer; C and N are presented as %, and all other nutrients are presented as ppm. About 3.42 and 1.71 gm of dried biofertilizer were added to 1 kg of soil for TB and HH treatments, respectively. Similar lowercase letters indicate no significant difference at $p < 0.10$.

On an economic aspect, we calculated that the application of plant nutrients through cyanobacteria biofertilizers will reduce the cost of production by 12 to 15% compared to using synthetic fertilizers. In addition, the use of cyanobacteria biofertilizers can fully or partially substitute chemical fertilizers without any loss of crop yield (okra). Therefore, cyanobacteria biofertilizers can help growers generate high-value organic okra and stay competitive in the market. South Florida soils are deficient in Fe (due to alkaline parent materials), and organic growers often face serious Fe deficiency challenges during crop production. Most of the growers apply chelated Fe as an organic Fe source, which substantially increases the cost of production. Cyanobacteria biofertilizers can provide 2000 ppm of Fe in the soil (Table 2), and substituting expensive organic Fe sources with cyanobacteria biofertilizers can significantly alleviate production costs for these growers.

Author Contributions: Conceptualization, S.C. and S.D.; Methodology, S.D. and S.C.; Formal analysis, S.C.; Resources, K.J.; Data curation, S.C.; Writing—original draft, S.D.; Writing—review & editing, S.D., S.C., K.J., L.J.S. and M.B.; Visualization, S.C., S.D., L.J.S. and M.B.; Funding acquisition, S.C., S.D. and K.J. All authors have read and agreed to the published version of the manuscript.

Funding: This research was funded by the Southern Sustainable Agriculture Research and Education (Southern SARE) R and E Grant # LS21-354.

Data Availability Statement: All data is available in the manuscript.

Acknowledgments: The authors would like to thank Tanaji Paul at FIU for helping us during setting up the experiments and collecting data. The company AECOM Inc., helped during slurry collection.

Conflicts of Interest: The authors declare no conflict of interest.

References

1. Chittora, D.; Meena, M.; Barupal, T.; Swapnil, P.; Sharma, K. Cyanobacteria as a source of biofertilizers for sustainable agriculture. *Biochem. Biophys. Rep.* **2020**, *22*, 100737. [CrossRef]
2. Simranjit, K.; Kanchan, A.; Prasanna, R.; Ranjan, K.; Ramakrishnan, B.; Singh, A.K.; Shivay, Y.S. Microbial inoculants as plant growth stimulating and soil nutrient availability enhancing options for cucumber under protected cultivation. *World J. Microbiol Biotechnol.* **2019**, *35*, 51. [CrossRef]
3. Manjunath, M.; Kanchan, A.; Ranjan, K.; Venkatachalam, S.; Prasanna, R.; Ramakrishnan, B.; Hossain, F.; Nain, L.; Shivay, Y.S. Rai, A.B.; et al. Beneficial cyanobacteria and eubacteria synergistically enhance bioavailability of soil nutrients and yield of okra *Heliyon* **2016**, *2*, e00066. [CrossRef] [PubMed]
4. Rossi, F.; Olguín, E.J.; Diels, L.; De Philippis, R. Microbial fixation of CO_2 in water bodies and in drylands to combat climate change, soil loss and desertification. *New Biotechnol.* **2015**, *32*, 109–120. [CrossRef]
5. Li, Y.C.; Klassen, W.; Lamberts, M.; Olczyk, T.; Liu, G. Okra Production in Miami-Dade County, Florida. HS-87/TR009. 2021 Available online: https://edis.ifas.ufl.edu/publication/TR009 (accessed on 30 August 2022).
6. Karthikeyan, N.; Prasanna, R.; Nain, L.; Kaushik, B.D. Evaluating the potential of plant growth promoting cyanobacteria as inoculants for wheat. *Eur. J. Soil Biol.* **2007**, *43*, 23–30. [CrossRef]
7. Prasanna, R.; Babu, S.; Rana, A.; Kabi, S.R.; Chaudhary, V.; Gupta, V.; Kumar, A.; Shivay, Y.S.; Nain, L.; Pal, R.K. Evaluating the establishment and agronomic proficiency of cyanobacterial consortia as organic options in wheat–rice cropping sequence. *Exp. Agric.* **2013**, *49*, 416–434. [CrossRef]
8. Jha, M.N.; Prasad, A.N. Efficacy of new inexpensive cyanobacterial biofertilizer including its shelf-life. *World J. Microbiol. Biotechnol.* **2006**, *22*, 73–79. [CrossRef]
9. Asmamaw, M.; Wolde, G.; Yohannes, M.; Yigrem, S.; Woldemeskel, E.; Chala, A.; Davis, J.G. Comparison of cyanobacterial bio-fertilizer with urea on three crops and two soils of Ethiopia. *Afr. J. Agric. Res.* **2019**, *14*, 588–596.
10. Chittapun, S.; Limbipichai, S.; Amnuaysin, N.; Boonkerd, R.; Charoensook, M. Effects of using cyanobacteria and fertilizer on growth and yield of rice, Pathum Thani I: A pot experiment. *J. Appl. Phycol.* **2018**, *30*, 79–85. [CrossRef]
11. Yoder, N.; Davis, J.G. Organic fertilizer comparison on growth and nutrient content of three kale cultivars. *HortTechnology* **2020**, *30*, 176–184. [CrossRef]
12. Faridmarandi, S.; Khare, Y.P.; Naja, G.M. Long-term regional nutrient contributions and in-lake water quality trends for Lake Okeechobee. *Lake Reserv. Manag.* **2020**, *37*, 77–94. [CrossRef]
13. Pereira, I.; Ortega, R.; Barrientos, L.; Moya, M.; Reyes, G.; Kramm, V. Development of a biofertilizer based on filamentous nitrogen-fixing cyanobacteria for rice crops in Chile. *J. Appl. Phycol.* **2009**, *21*, 135–144. [CrossRef]
14. Romanowska-Duda, Z.; Szufa, S.; Grzesik, M.; Piotrowski, K.; Janas, R. The Promotive Effect of *Cyanobacteria* and *Chlorella* sp Foliar Biofertilization on Growth and Metabolic Activities of Willow (*Salix viminalis* L.) Plants as Feedstock Production, Solid Biofuel and Biochar as C Carrier for Fertilizers via Torrefaction Process. *Energies* **2021**, *14*, 5262.
15. Dattamudi, S.; Wang, J.J.; Dodla, S.K.; Arceneaux, A.; Viator, H.P. Effect of nitrogen fertilization and residue management practices on ammonia emissions from subtropical sugarcane production. *Atmos. Environ.* **2016**, *139*, 122–130. [CrossRef]
16. Xiong, D.; Chen, J.; Yu, T.; Gao, W.; Ling, X.; Li, Y.; Peng, S.; Huang, J. SPAD-based leaf nitrogen estimation is impacted by environmental factors and crop leaf characteristics. *Sci. Rep.* **2015**, *5*, 13389. [CrossRef]
17. Garcia, C.L.; Dattamudi, S.; Chanda, S.; Jayachandran, K. Effect of salinity stress and microbial inoculations on glomalin production and plant growth parameters of snap bean (*Phaseolus vulgaris*). *Agronomy* **2019**, *9*, 545. [CrossRef]

18. Kandel, B.P. Spad value varies with age and leaf of maize plant and its relationship with grain yield. *BMC Res. Notes* **2020**, *13*, 475. [CrossRef]
19. Meng, Z.; Duan, A.; Chen, D.; Dassanayake, K.B.; Wang, X.; Liu, Z.; Gao, S.; Liu, H. Suitable indicators using stem diameter variation-derived indices to monitor the water status of greenhouse tomato plants. *PLoS ONE* **2017**, *12*, e0171423. [CrossRef]
20. Clough, A.; Hunter, M.N. Stem diameter: A rapid accurate parameter for monitoring growth of sorghum. In Proceedings of the 11th Australian Agronomy Conference, Geelong, Victoria, 2–6 February 2003; Volume 22, p. 2005.
21. Toribio, A.J.; Suárez-Estrella, F.; Jurado, M.M.; López, M.J.; López-González, J.A.; Moreno, J. Prospection of cyanobacteria producing bioactive substances and their application as potential phytostimulating agents. *Biotechnol. Rep.* **2020**, *26*, e00449. [CrossRef]
22. Essa, A.M.; Ibrahim, W.M.; Mahmud, R.M.; ElKassim, N.A. Potential impact of cyanobacterial exudates on seed germination and antioxidant enzymes of crop plant seedlings. *Int. J. Curr. Microbiol. Appl. Sci.* **2015**, *4*, 1010–1024.
23. Agwa, O.K.; Ogugbue, C.J.; Williams, E.E. Research Article Field Evidence of Chlorella vulgaris Potentials as a Biofertilizer for Hibiscus esculentus. *Int. J. Agric. Res.* **2017**, *12*, 181–189. [CrossRef]

environments

MDPI

Article

Reuse of Soils Fertilized with Ash as Recycling Derived Fertilizer Revealed Strong Stimulation of Microbial Communities Involved in P Mobilization in *Lolium perenne* Rhizospheres

Lea Deinert and Achim Schmalenberger *

Department of Biological Sciences, University of Limerick, V94 T9PX Limerick, Ireland
* Correspondence: achim.schmalenberger@ul.ie; Tel.: +353-61-233775

Abstract: Circular economy recycling-derived fertilizers (RDF) have the potential to replace linear economy fertilizers such as unsustainable superphosphates. Here, effects of ash RDF treatments in Irish grassland cultivation were investigated in a simulated second growing season. Soil fertilized in a preceding pot trial with superphosphate (SP), poultry-litter ash (PLA) and sewage-sludge ash (SSA) at P concentration of 60 kg P ha^{-1} and a P-free control (SP0) was reused in a microcosm trial. *Lolium perenne* was cultivated for 54 days in six replicates with a full complement of micro- and macro-nutrients other than P. PLA treatments provided higher dry weight shoot yields than SP0, while SSA and SP overlapped with SP0 and PLA. Most probable number (MPN) analysis showed that phosphonate- and phytate-utilizing bacterial abundance was significantly increased in PLA. Alkaline (*phoD*) phosphomonoesterase gene fragments were significantly more abundant (qPCR) in the ashes than the superphosphate or P-free control. Bacterial communities were significantly affected by the P application. Similarly, a significant separation of treatments was confirmed in a canonical correspondence analysis of the *phoD*-harboring community. The genera *Streptomyces* and *Xanthomonas* were significantly higher in abundance in the ash RDFs. These results demonstrated the potential benefits of ash RDF treatments as an alternative P source.

Keywords: ryegrass; circular economy; bacterial community structure; *phoD*; *phoC*; phosphatase; next generation sequencing; DGGE; qPCR

1. Introduction

Fertilizer supply plays an important role in securing food production for the growing human population. The United Nations predicts an increase in the global population from 7.7 billion in 2019 to between 9.4 to 10.1 billion in 2050 [1]. Conventional mineral P fertilizer is produced from phosphate rock, a non-renewable resource located in a small number of mines around the world, with the highest abundance in geopolitically unstable countries. Annually, around 22×10^{12} g mined fossil P enters the global economy, and mined P is still the source for 80% of P fertilizers produced for agricultural purposes [2,3]. Mining of phosphate rock occurs at a significantly faster rate than it can replenish itself via natural geochemical processes, that takes 10–15 million years [4,5]. This leads to the depletion of the finite resource in a linear nutrient economy approach ("take, make, dispose"). Simultaneously it pollutes the environment with a surplus of unequally distributed nutrients originating from the intensification of agriculture and subsequent P accumulation in soil causing run-off and leaching of P. Nutrient recovery from waste streams is a promising move towards a circular economy ("reduce, recycle, reuse") [6,7]. The recovered nutrients can be applied to soil as recycling-derived fertilizers (RDFs), endeavoring to close the loop in the P cycle.

In soil, P is a relatively immobile nutrient. It is present in various organic and inorganic forms in different pools, which represent occluded/stable, labile, to readily available P fractions of soil [8]. Inorganic and organic P can also be sorbed to soil particles and become unavailable as a source of nutrient [9,10]. Plants almost exclusively access P in the form of orthophosphate, which is typically present in low concentrations between 0.1 to 1 mg P kg^{-1} in the soil solution, making up around 1% of soil total P [11]. It was estimated that only up to 30% of mineral P fertilizer applied to arable land was taken up by plants, with the remaining 70% either being immobilized or sorbed via microbial or physicochemical processes or leached into the environment [12,13]. The soil microbiota therefore plays a crucial role in accessing P and transforming it into plant available forms. There are a multitude of mechanisms improving P plant uptake facilitated by microorganisms, including: (a) extension of the root system via symbiotic relationships with arbuscular mycorrhizal fungi (AMF), the AMF hyphae can penetrate smaller soil pores compared to the plant's root hairs and thus explore larger soil volumes; (b) P sorption equilibria alterations in favor of net flux of orthophosphate into the soil solution; (c) direct or indirect microbial P turnover that includes proton efflux, phosphatase production and extracellular release, siderophore exudation, etc. [14,15]. In recent years, soil degradation, acidification and loss of biodiversity have been reported in agriculturally managed soils [16]. This includes a reduction of the alpha diversity of AMF—important plant symbionts for improved P acquisition by the plant host—that was reported in response to the application of mineral fertilizers due to the direct P availability for plant uptake, while organic fertilizers reinforced the plant-fungal symbiosis [17]. Likewise, orthophosphate application in grassland columns negatively affected the alkaline phosphatase gene, nematode and AMF abundance and shifted microbial communities [18]. Application of ashes, that were left over from energy and heat production processes, as fertilizers has been practiced in the past [19]. Ashes usually increase soil pH, potentially counteracting soil acidification and the liming effect also appears to increase bacterial activity [20]. Due to the heat treatment, ashes lack N, and are therefore not feasible for application on mineral soils with low N index [21]. While increased microbial activity after ash application implies carbon release from soil due to higher soil respiration [22], other microbial processes involved in nutrient cycling and therefore P mobilization might be enhanced. This mobilization of P may include the mineralization of organically bound P from the soil already present prior to the fertilizer application as well as recently immobilized P of fertilizer origin. Depending on the feedstock of ash fertilizers (i.e., sewage sludge), a high heavy metal burden is possible, that may accumulate in soil [23]. Therefore, the development of RDFs should focus both on the plant availability of P as well as the removal of hazardous substances from waste streams. The AshDec® process was developed for the incineration of sewage sludge, whereby potassium or sodium based alkaline products are added to the feedstock during incineration at temperatures above 900 °C, to obtain ammonium citrate soluble, alkaline P compounds, indicating their plant availability, and simultaneously evaporating toxic heavy metals such as arsenic, cadmium and lead from the product [24].

The objective of this study was to determine (a) the P plant availability; (b) the microbial-mediated P mineralization activity in experimentally re-used soil after application of two ash RDF fertilizers at a rate of 60 kg P ha^{-1} in the preceding year and one cut of *Lolium perenne* (perennial ryegrass) grass; and (c) compare *L. perenne* shoot production and AMF colonization. Feedstock for ash production was poultry litter (PLA) and municipal sewage sludge (SSA), in comparison to a P-free control (SP0) and conventional superphosphate (SP) fertilizer. RDF based plant growth promotion alongside mineral fertilizer use with and without superphosphate was correlated to changes in the bacterial (16S rRNA gene and *phoD*) community structures in rhizosphere soil, bacterial phosphonate and phytate utilization, potential phosphatase activity and arbuscular mycorrhizal root colonization. The hypothesis was that P supplied from ash RDFs, may have a lower detrimental impact on the P mobilizing microbial community than mineral P fertilizers and promotes the abundance of P mobilizing bacteria and AMF interactions in the longer term.

2. Materials and Methods

2.1. Microcosm Trial Set-Up

The bulk soil from the preceding pot trial experiment [25], stored in sealed plastic bags in the dark at ambient temperatures was reutilized for this experiment. The soil originated from a field at Teagasc Johnstown Castle, Wexford, Ireland (N52°17′47″, W6°30′29″), which had been reseeded with perennial ryegrass in 2018 and prior to that was under long-term pasture management. The P-free control SP0, the mineral fertilizer treatment SP60 and the two ashes PLA60 and SSA60 at 60 kg P ha^{-1} were chosen from the preceding pot trial (see also Supplementary Information S1). Microcosms (chemically welded plexiglass, length = 12 cm, height = 11 cm, width = 2.5 cm; [26]) were set up in replicates of six per treatment. The moisture content of each treatment was determined via the oven dry method at 105 °C for 24 h. Water holding capacity of the soil was measured for each treatment prior to the start of the microcosm experiment. First, 330 g of soil was used per microcosm and the soil was watered to reach 70% water holding capacity. Then, 0.02 g of the diploid *Lolium perenne* (var. AberGreen) seeds were added on the soil surface. The seeds were covered with an additional 30 g of soil of the same treatment and that layer was wetted evenly with additional 5 mL of distilled water. The microcosms were incubated in a plant growth chamber (A1000, Conviron, Manitoba, Canada), with the following conditions: 12 h day and 12 h night cycle, the photosynthetically active radiation (400–700 nm) was set between 200–400 µmol ms^{-1}, at a temperature and relative humidity of 20 ± 2 °C and 75 ± 5% r.H. during the photoperiod and 15 ± 1 °C and 70 ± 5% r.H. during darkness. Until germination was observed, the microcosms were kept in propagators to minimise moisture losses. The plexiglass microcosms were wrapped in aluminium foil to minimise algal growth. The microcosms were irrigated with distilled water every second day, at the beginning with a volume of 5 mL, which was then increased to 25 mL over time to meet the plants' needs. To prevent N-starvation effect in this experiment, a urea solution was applied at a rate of 75 kg N ha^{-1} with a volume of 10 mL per microcosm.

2.2. Microcosm Harvest, Plant and Soil Analyses

After 54 days of perennial ryegrass cultivation in the plant growth chamber, the bulk soil was harvested separately from the rhizosphere soil. The soil-root mass was transferred into plastic bags and the loosely attached bulk soil was shaken off. Then, the roots were transferred into a zip-lock bag, where the rhizosphere soil, which is closely attached to the root surface, was largely removed by vigorous shaking, until no more soil would detach from the roots. Roots were picked from each sample for arbuscular mycorrhizae analysis, together with the bulk and rhizosphere soil, and stored at 4 °C. Subsamples of the rhizosphere soil were taken and stored at −20 °C for DNA extraction. The number of tillers was counted for each microcosm and the fresh weight of the plant shoots was recorded. The plant biomass was dried in an oven at 55 °C for 72 h, then the dry weight of the plant yield was determined.

Soil pH, available P, potential soil acid and alkaline phosphomonoesterase activity (ACP, ALP) were determined described previously [25]. In brief, soil pH was determined in 0.01M $CaCl_2$ solution. Morgan's P test was set up to determine plant availability of P following standard protocols [25,27]. Soil's potential acid and alkaline phosphomonoesterase (phosphatase) activity was measured in 1 g of soil spectrophotometrically with the p-nitrophenyl phosphate method following standard protocols [28].

2.3. Cultivation Dependent Microbial Analyses

The analyses MPN determination in minimal media with a single P source were conducted as described previously [25] for the pot trial. Briefly, for MPN a serial dilution of rhizosphere soil was carried out in saline (0.85% *w/v*). For MPN, 96 well microtiter plates were used with minimal medium containing either phosphonoacetic acid (MM2PAA) or phytate (MM2Phy) as sole source of P as described elsewhere [29] and incubated at 25 °C for 14 days to visually determine growth. An MPN approach was chosen over plate-based assays to eliminate traces of P commonly found in solidifying agents. MPN g^{-1} rhizosphere soil was determined via published tables by the Food and Drug Administration (FDA, Blodgett, https://www.fda.gov/food/laboratory-methods-food/bam-appendix-2-most-probable-number-serial-dilutions, accessed on 16 June 2022).

2.4. Cultivation Independent Microbial Analyses

Arbuscular mycorrhization was analysed according to protocols as described elsewhere [18,30]. Briefly, 10 cm long root pieces were incubated in a 10% KOH solution in a water bath at 90 °C for 45 min. Then the roots were rinsed three times with distilled water, before they were incubated with freshly prepared alkaline 30% hydrogen peroxide at room temperature for 60 min. After that, the roots were rinsed with distilled water again and covered with 10 mL 0.1 M hydrochloric acid overnight for 12 h. The next day, the acid was discarded, and the roots were stained with a 0.05% trypan blue lactoglycerol solution at 90 °C for 45 min. Finally, the staining solution was discarded, and the roots were stored in a lactoglycerol discoloration solution until analysis was carried out. Root colonization with arbuscular, vesicular and hyphal structures was evaluated with an Olympus BX60 microscope (Tokyo, Japan) at 600x magnification in an adapted gridline-intersect method [30]. Four 1 cm long root pieces were investigated per sample and for each piece of root 25 fields of view were observed and the presence of AMF structures was recorded (Supplementary Figure S1) and calculated in percentage.

16S PCR-DGGE, *phoC* and *phoD* qPCR and *phoD* sample preparation for sequencing were conducted as described previously [25] for the pot trial. In brief, DNA extraction from 0.25 g rhizosphere soil (frozen at −20 °C) was performed following the manufacturer's instructions of the DNeasy PowerSoil Pro kit (QIAGEN GmbH, Hilden, Germany). DNA was quantified with the Qubit Fluorometer (Life Technologies, Carlsbad, CA, USA) using a Qubit dsDNA HS assay kit (Life Technologies, Carlsbad, CA, USA).

In order to provide an overview of the bacterial community structures, 16S rRNA gene PCR for denaturing gradient gel electrophoresis (DGGE) was carried out with primer pair 341F-GC and 518R [31]. Conditions for DGGE are provided elsewhere [25]. Amplicon based sequencing of the V4 region of the 16S rRNA gene was conducted with primers containing Illumina adapters (515F-Illumina 5′- TCG TCG GCA GCG TCA GAT GTG TAT AAG AGA CAG GTG CCA GCM GCC GCG GTA A-3′ and 806R-Illumina 5′- GTC TCG TGG GCT CGG AGA TGT GTA TAA GAG ACA GGG ACT ACH VGG GTW TCT AAT-3′) using published PCR protocols [25] in order to obtain sequence based information that goes beyond the capabilities of a finger-print based analysis. Next-Generation Sequencing was performed at the University of Minnesota Genomics Center (Minneapolis, MN, USA) on a MiSeq PE300 platform. Sequencing of the *phoD* gene was conducted by amplifying the *phoD* gene first with primers phoD-F733 (5′-TGG GAY GAT CAY GAR GT-3′) and phoD-R1083 (5′-CTG SGC SAK SAC RTT CCA-3′) [32]. Amplicons were subjected to a second PCR in order to add the Illumina adapter sequences (Adapter F: TCG TCG GCA GCG TCA GAT GTG TAT AAG AGA CAG; Adapter R: GTC TCG TGG GCT CGG AGA TGT GTA TAA GAG ACA G). The PCR conditions for first and second reaction have been published elsewhere [25]. Amplicon sequence reads (16S rRNA gene fragments and *phoD* gene fragment) have been deposited with links to BioProject accession number PRJNA1043576 in the NCBI BioProject database (https://www.ncbi.nlm.nih.gov/bioproject/, accessed on 23 November 2023; SAM38345078-101 (16S), SAMN38345078-101 (*phoD*)).

qPCR was conducted for *phoC* with the primers phoC-A-F1 (5′-CGG CTC CTA TCC GTC CGG-3′) and phoC-A-R1 (5′-CAA CAT CGC TTT GCC AGT G-3′) [33] as well as for *phoD* with ALPS-F730 (5′-CAG TGG GAC GAC CAC GAG GT-3′) and ALPS-R1101 (5′-GAG GCC GAT CGG CAT GTC G-3′) [34] on a Roche LightCycler® 96 (Roche Diagnostics, Mannheim, Germany) using the KAPA SYBR FAST qPCR Master Mix (KAPA Biosystems, Cape Town, South Africa) as described previously [25].

2.5. Data Analysis

Shoot weights, pH, MPN, CFU, ACP, ALP, arbuscular (AC), vesicular (VC) and hyphal (HC) colonization, Morgan's P, *phoD* and *phoC* absolute quantification were assessed for significant differences with SPSS (SPSS Statistics, IBM, Version 26). Normality was tested with Shapiro–Wilk ($p > 0.05$). The homogeneity of variance was evaluated using Levene's test at $p > 0.05$. When both criteria were fulfilled, a one-way analysis of variance (ANOVA) was performed (Tukey HSD post-hoc, $p \leq 0.05$). Data violating assumptions of normality were transformed (untransformed results were reported), a non-parametric Kruskal–Wallis test was performed when normality was not achieved. When equal variance was not fulfilled a Games–Howell post-hoc test was conducted.

DGGE gel images were analysed with Phoretix 1D (Totallab, Newcastle, UK) to create binary matrixes for canonical correspondence analysis (CCA) with all environmental data described above using the R packages vegan, mabund, permute and lattice, applying a permutational multivariate analysis of variance (PERMANOVA) test using 999 permutations (R Studio, Version 4.0.3).

Demultiplexed 16S rRNA gene amplicon sequences returned from University of Minnesota Genomics Center (Minneapolis) were analysed in QIIME2 2020.8 [35] using the q2-dada2 plugin (paired-end reads joined, quality filtered, denoised) [36]. Obtained amplicon sequence variants (ASV) were aligned (q2-alignment with mafft [37]) and a phylogenetic tree was built (q2-phylogeny with fasttree2 [38]). Alpha [39] and beta diversity metrics (weighted UniFrac [40]) were created using the q2-diversity plugin and taxonomy assignments were conducted (q2-feature-classifier [41], SILVA 13.8 99% reference data set for 16S rRNA [42,43]). References were trimmed to train a Naïve Bayes classifier. R (Version 4.0.3) with mvabund and vegan packages were used to carry out CCA analysis. Taxa bar plots were created to display relative abundance distributions of prevalent taxa. Differential abundance was tested using Kruskal–Wallis rank sum test (kruskal.test) with Wilcoxon post-hoc analysis for multiple pairwise comparisons between groups (function pairwise.wilcox.test) and Benjamini–Hochberg correction.

Sequences of *phoD* were trimmed of primers and poor-quality sequences (cutadapt [44]). Paired end reads were merged (usearch [45]) and filtered (fastq_filter). Unique and duplicate sequences were removed (fastx_uniques command). UPARSE was used to cluster sequences into centroid operational taxonomic units (OTUs) [46] at a 75% sequence similarity threshold [47]. An alkaline phosphatase gene reference database (fungene [48]) was used to assign taxonomy. CCA biplot, bar plots and subsequent permutation analysis was conducted in R as described above.

3. Results

3.1. Biomass Yields, Agronomic Efficiency and Elemental Analysis

The fresh weight of the *L. perenne* shoots after the microcosm harvest was highest in the SP60 treatment, while the no P control SP0 was the lowest (Table 1). The dry biomass analysis showed that both RDF treatments had higher yields compared to the mineral SP60 treatment with PLA60 being significantly higher than SP0. However, after correcting for the seed germination rate (average dry weight per shoot) there was no significant difference detectable between any of the treatments (Table 1). Agronomic efficiency (P-uptake/P-supplied) was highest for PLA60 and lowest for SP60 (Table 1).

The mass balance for the macro- and micronutrients from pooled plant dry matter per pot (i.e., nutrient uptake) identified lowest P and N levels in PLA60 (Table 2). Interestingly, K content was around 20% higher in SP0 compared to all other treatments. SP60 had the highest Ca, Mg, N, P and S content compared to the ash fertilized treatments. Concerning the micronutrients, B and Mb were similar for all treatments, while Cu and Fe uptake were lowest in the PLA60 treatment. Mn mass balance was highest in SP0, while Zn was lower in both ash treatments.

Table 1. Average values for fresh weight (FW) and dry weight (DW) of *Lolium perenne* harvest, shoot count, dry weight to shoot ratio, and agronomic efficiency (AE).

Treatment	Biomass FW (g)		Biomass DW (g)		Shoot Count		DW:Shoot Ratio (g Shoot^{-1})		AE
SP0	11.87 ^{b}	±0.33	1.49 ^{b}	±0.16	87.3 ^{a}	±2.73	0.017 ^{a}	$\pm 1.7 \times 10^{-03}$	
SP60	14.63 ^{a}	±0.31	1.71 ^{ab}	±0.05	87.2 ^{a}	±3.72	0.020 ^{a}	$\pm 7.8 \times 10^{-04}$	13.5
PLA60	13.13 ^{ab}	±0.24	1.84 ^{a}	±0.02	86.3 ^{a}	±2.89	0.021 ^{a}	$\pm 5.5 \times 10^{-04}$	20.8
SSA60	12.76 ^{b}	±0.71	1.77 ^{ab}	±0.04	84.5 ^{a}	±3.12	0.021 ^{a}	$\pm 1.1 \times 10^{-03}$	17.0

54 d of growth, n = 6, letters a,b indicate significant difference per column, ± = standard error.

Table 2. Mass balance of total perennial ryegrass shoot dry weight (pooled) per treatment after 54 days of growth.

Treatment	Macronutrients [µg]					
	Ca	K	Mg	N	P	S
SP0	1379	5135	325	4070	157	291
SP60	1381	4290	360	3911	214	292
PLA60	1025	4238	263	3094	136	245
SSA60	1157	4036	282	3565	151	263
Treatment	**Micronutrients [µg]**					
	B	**Cu**	**Fe**	**Mb**	**Mn**	**Zn**
SP0	0.75	1.32	35.20	0.01	22.19	4.63
SP60	0.67	1.28	21.21	0.02	16.86	4.58
PLA60	0.61	0.93	17.79	0.02	16.81	3.87
SSA60	0.58	1.02	24.74	0.01	16.03	3.95

Calcium (Ca), potassium (K), magnesium (Mg), nitrogen (N), phosphorus (P), sulphur (S), boron (B), copper (Cu), iron (Fe), molybdenum (Mb), manganese (Mn), zinc (Zn).

3.2. Microbial P-Mobilization and Solubilization, Soil pH and Morgan's P

Microbial P mobilization detected with the cultivation-dependent MPN approach demonstrated a significantly higher ($p \leq 0.05$) P mobilization capability from phosphonates (PAA) and phytate (Phy) in the PLA60 ash treatment (Table 3). SSA60 had an average of 3.7×10^6 MPN g^{-1} soil in the PAA medium, which exceeded that of PLA60 with 3.2×10^6 MPN g^{-1} soil, although significant difference was not reached. MPN PAA and Phy values for SP0 and SP60 were lower than the ash treatments but significant difference was only reached with PLA60. The overall number of heterotrophic bacteria able to grow in R2A medium was similar in all treatments.

Table 3. MPN values of phosphonoacetic acid utilizing (PAA) and phytate utilizing (Phy) bacteria, and MPN values of total heterotrophic bacteria (R2A) in rhizosphere soil of *L. perenne*.

Treatment	PAA (MPN g^{-1} Soil)		Phy (MPN g^{-1} Soil)		R2A (MPN g^{-1} Soil)	
SP0	6.2×10^{05} ^{b}	$\pm 2.0 \times 10^{05}$	3.5×10^{06} ^{b}	$\pm 7.2 \times 10^{05}$	2.0×10^{08} ^{a}	$\pm 1.4 \times 10^{08}$
SP60	1.6×10^{06} ^{b}	$\pm 1.1 \times 10^{06}$	6.7×10^{06} ^{b}	$\pm 1.3 \times 10^{06}$	8.7×10^{07} ^{a}	$\pm 2.1 \times 10^{07}$
PLA60	3.2×10^{06} ^{a}	$\pm 4.2 \times 10^{05}$	1.8×10^{07} ^{a}	$\pm 5.1 \times 10^{06}$	1.1×10^{08} ^{a}	$\pm 4.7 \times 10^{07}$
SSA60	3.7×10^{06} ^{ab}	$\pm 1.1 \times 10^{06}$	1.0×10^{07} ^{ab}	$\pm 3.0 \times 10^{06}$	7.8×10^{07} ^{a}	$\pm 1.1 \times 10^{07}$

54 d of growth, n = 6, letters a,b indicate significant difference per column, ± = standard error.

The soil pH was affected by all fertilizer treatments, all fertilizers showed a significant liming effect ($p \leq 0.05$) compared to the SP0 control, and the pH of the ash RDFs was again significantly higher than in the mineral SP60 treatment (Table 4). Moreover, the concentration of orthophosphates readily available for plant uptake, determined using the Morgan's P extraction, was significantly higher in all P fertilized treatments, with the SP60 and PLA60 treatment falling into a soil P index of 2, increasing the soil P status from "very low" to "low" compared to the SP0 and SSA60 treatment. The Morgan's P values for SP60 and PLA60 were twice as high as the one for the P-free SP0 control (Table 4).

Table 4. Mean values for soil pH (bulk soil) and plant available P (Morgan's; rhizosphere soil from *L. perenne* in microcosms.

Treatment	Soil pH		Morgan's P (mg P L^{-1})	
SP0	4.75 ^{c}	±0.01	1.53 ^{b}	±0.11
SP60	5.06 ^{b}	±0.03	3.12 ^{a}	±0.15
PLA60	5.34 ^{a}	±0.03	3.20 ^{a}	±0.55
SSA60	5.30 ^{a}	±0.04	2.43 ^{a}	±0.16

54 d of growth, *n* = 6, letters a–c indicate significant difference per column, ± = standard error.

3.3. Phosphatase Activity and Gene Copy Numbers (phoC, phoD) in Soil

The potential ACP activity was around ten-fold higher than for ALP, with PLA60 displaying the highest value on average for ACP and SP0 the highest average value for ALP (Table 5). However, neither ACP nor ALP activity in soil identified any significant differences ($p > 0.05$). Assessment of the gene copy numbers for the alkaline and acid phosphatases on the other hand revealed a significantly higher ($p \leq 0.05$) *phoC* copy number g^{-1} soil in the SSA treatment and a significantly higher *phoD* copy number g^{-1} soil in PLA60 and SSA60 (Table 5), which had also demonstrated significantly higher soil pH levels (Table 4).

Table 5. Potential acid (ACP) and alkaline phosphatase (ALP) and absolute *phoC* and *phoD* gene copy numbers per gram of *L. perenne* rhizosphere soil in microcosms.

Treatment	ACP (µg pNP g^{-1} Soil h^{-1})		ALP (µg pNP g^{-1} Soil h^{-1})		*phoC* (*phoC* Copies g^{-1} Soil)		*phoD* (*phoD* Copies g^{-1} Soil)	
SP0	1504.9 ^{a}	±79.0	121.9 ^{a}	±11.1	1.1×10^{07} ^{b}	$\pm 1.7 \times 10^{06}$	1.5×10^{09} ^{b}	$\pm 1.8 \times 10^{08}$
SP60	1556.1 ^{a}	±65.2	109.7 ^{a}	±8.3	1.1×10^{07} ^{b}	$\pm 9.7 \times 10^{05}$	1.5×10^{09} ^{b}	$\pm 5.8 \times 10^{07}$
PLA60	1608.1 ^{a}	±91.4	117.4 ^{a}	±6.5	2.2×10^{07} ^{ab}	$\pm 3.3 \times 10^{06}$	2.8×10^{09} ^{a}	$\pm 2.7 \times 10^{08}$
SSA60	1322.1 ^{a}	±94.0	115.9 ^{a}	±6.7	2.2×10^{07} ^{a}	$\pm 8.0 \times 10^{05}$	2.5×10^{09} ^{a}	$\pm 1.1 \times 10^{08}$

54 d of growth, *n* = 6, letters a,b indicate significant difference per column, ± = standard error.

3.4. Arbuscular Mycorrhizal Fungi Root Colonization

Root colonization with AMF was assessed via root staining and subsequent microscopic evaluation (Supplementary Table S1). There were no statistically significant differences ($p > 0.05$) detected between the treatments as arbuscular colonization ranged between 43 and 63%, vesicular colonization ranged from 53 to 76% and hyphal colonization exceeded 90%. Nevertheless, SP60 had the lowest average colonization with arbuscules and hyphae, compared to other treatments.

3.5. Bacterial Community Analysis

The 16S PCR-DGGE analysis of the bacterial community in the rhizosphere soil of the microcosm trial resulted in significant separation of the treatments ($p < 0.01$), as shown in the CCA biplot in Figure 1. The treatments SP0, SP60 and PLA60 were separated from SSA60 on the first axis, and PLA60 was separated from all other treatments on the second axis. Pairwise comparison via PERMANOVA of all treatments using Benjamini–Hochberg correction confirmed significant differences ($p \leq 0.05$) in the bacterial community for all treatments. Significant parameters ($p \leq 0.05$) involved in the shift of the bacterial communities in the treatments SSA60 and PLA60 were plant available P, shoot dry weight yield, soil pH, phosphonate and phytate utilizing bacteria, *phoC* and *phoD* copy number and vesicular root colonization.

For the 16S rRNA amplicon NGS analysis of DNA extracted from the rhizosphere soil of the microcosm trial, a total of 4,195,879 demultiplexed sequences were received from the sequencing facility. After filtering, merging and chimera detection using DADA2, an average of 51,380 features (29.5%) per sample were obtained for further analysis in the QIIME2 platform. None of the alpha diversity estimators were statistically significantly different (Supplementary Table S2). Shannon index ranged from 5.8 to 6.2, while Simpson index exceeded 0.995.

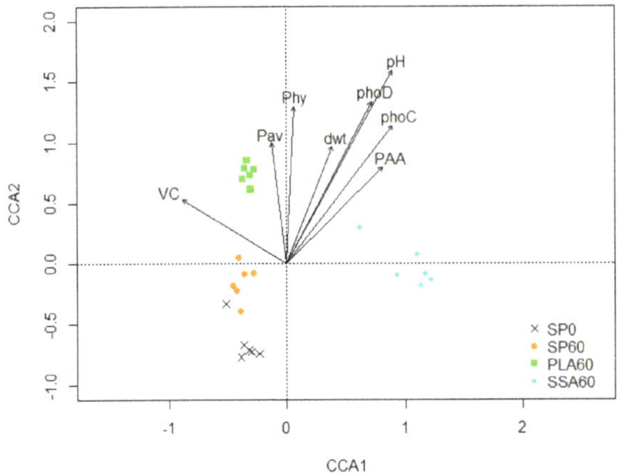

Figure 1. 16S rRNA gene PCR-DGGE based CCA biplot with environmental parameters plant dry weight (dwt), *phoC* and *phoD* copy numbers, soil pH, phosphonate (PAA) and phytate (Phy) utilizing MPN counts, plant available P (Pav) and AMF vesicle colonization (VC), significantly ($p < 0.05$) affecting the rhizosphere soil bacterial community; CCA1 explains 30.02% and CCA2 14.16% of the total variation of the data, $n = 6$.

A CCA biplot of the ASV matrix obtained from the 16S rRNA amplicon sequencing results (Figure 2) revealed clear separations of all four treatments. The treatments SP0 and SP60 are separated from the two ash RDF treatments on the first axis, and SP60 and SSA60 are separated from SP0 and PLA60 on the second axis. A pairwise comparison using PERMANOVA confirmed significant separation ($p \leq 0.05$) of all treatments after Benjamini–Hochberg correction for multiple testing. Canonical analysis environmental variable analysis revealed that vesicle colonization, P-mobilization capability from phytate and phosphonoacetic acid, bioavailable P, shoot dry weight, soil pH and *phoC* and *phoD* copy number significantly affected the bacterial community structures ($p \leq 0.05$).

The ten most abundant phyla represented over 90% of the overall phyla present (Supplementary Figure S2). In decreasing relative abundance these were *Actinobacteria*, *Firmicutes*, *Proteobacteria*, *Planctomycetes*, *Verrucomicrobia*, *Chloroflexi*, *Acidobacteria*, *Patescibacteria*, *Bacteriodetes* and *Thaumarchaeota*. Post-hoc analyses of their relative abundance between treatments revealed significant differences ($p \leq 0.05$) in all but two of the ten most abundant phyla. *Actinobacteria* and *Bacteroidetes* were the only of the top 10 phyla that were not significantly different in abundance ($p > 0.05$) across the four treatments (Figure 3). While the phyla *Firmicutes* and *Patescibacteria* were significantly relative lower in abundance in the two ash RDF treatments when compared to SP0 and SP60, the opposite was observed for the phyla *Chloroflexi* and *Planctomycetes*. *Acidobacteria* and *Thaumarchaeota* were significantly relative higher in abundance in PLA60 than all other treatments, and still significantly relative higher in abundance in SSA60 than in the SP0 treatment. *Verrucomicrobia* were significantly relative higher in abundance in SP60 as well as PLA60. *Proteobacteria* was the only phylum which had a significantly higher relative abundance in the SSA60 treatment. At genus level, no significant differences between the treatments had been detected for the bacterial community.

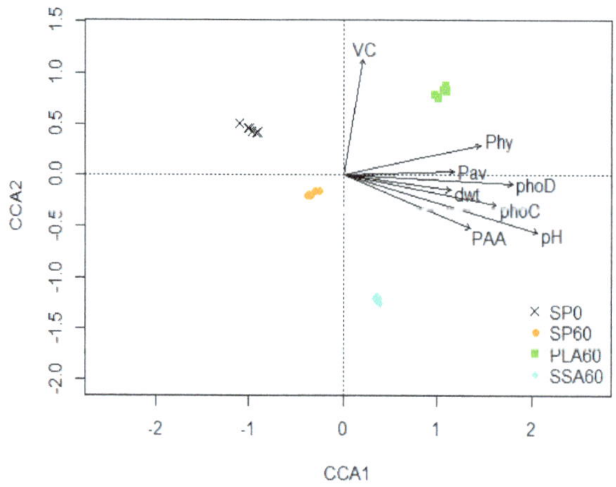

Figure 2. CCA plot of the 16S rRNA amplicon sequencing data of the microcosm trial, significantly correlated environmental parameters are perennial ryegrass dry weight (dwt), bioavailable P determined via Morgan's P (Pav), phosphonate-P mobilizing bacteria (PAA), phytate–P mobilizing bacteria (Phy), acid (*phoC*) and alkaline phosphatase (*phoD*) gene copy numbers, soil pH (pH), vesicular root colonisation by AMF (VC), CCA1 explains 5.03% and CCA2 4.64% of the total variation of the data, $n = 6$.

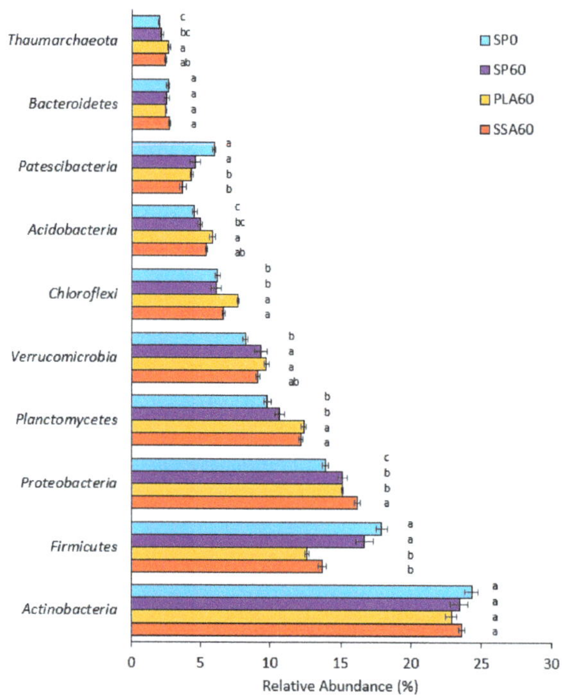

Figure 3. Horizontal bar plot of top 10 relative abundance phyla for SP0 (blue, 1st top), SP60 (purple, 2nd top), PLA60 (yellow, 3rd top) and SSA60 (red, bottom) (16S rRNA NGS data), error bars state standard error, significance determined via Kruskal–Wallis test and Wilcoxon post-hoc analysis and Benjamini–Hochberg correction, different letters a–c indicate statistically significant differences for each phylum ($p \leq 0.05$), $n = 6$.

3.6. Bacterial Community Analysis Based on phoD

There were 3,169,711 *phoD* functional gene paired-end sequences obtained from the sequencing facility, out of which 1,934,508 (60.9%) were successfully merged in USEARCH. Of the merged sequences, 612,338 were matched with zero differences. In total, 1,285,676 (66.6%) of the merged number of sequences passed the filter criteria and 765,090 (88.4%) of those were singletons. In total, 5051 OTUs were picked at a 75% similarity threshold and 2476 chimeral sequences (2.5%) were removed. Finally, 1,449,560 sequences were mapped to OTUs. The alpha diversity index Simpson was significantly lower for the PLA60 treatment (0.987 vs. 0.988–0.989) compared to all other treatments, and all other alpha diversity estimators (i.e., Shannon, ACE, Chao1) showed no significant differences between the treatments (Supplementary Table S3). The *phoD* amplicon NGS data based CCA biplot showed significant separation ($p \leq 0.05$) of all treatments (Figure 4; PERMANOVA, Adonis, confirmed via Benjamini–Hochberg correction). PLA60 and SSA60 were separated from the SP0 and SP60 treatment on the first axis, and on the second axis a separation between PLA60 and SSA60 is visible. Various environmental parameters were significantly affecting the *phoD* harboring community that included available P, *L. perenne* dry weight, soil pH, *phoC* and *phoD* gene copy numbers, phosphonate and phytate utilizing bacteria (MPN).

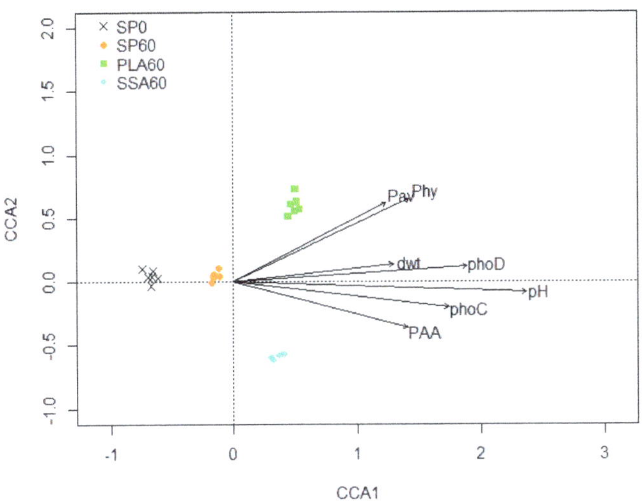

Figure 4. CCA biplot of the *phoD* bacterial community structure of the microcosm trial, CCA1 explains 6.68% and CCA2 4.49% of the total variation of the data environmental parameters; plant available P (Pav), phytate (Phy) and phosphonate (PAA) utilizing bacteria, *L. perenne* shoot dry weight (dwt), *phoC* and *phoD* copy numbers and soil pH affected the community structure significantly ($p < 0.05$), $n = 6$.

The ten most abundant genera detected in the analysis of the *phoD* sequences were, with decreasing relative abundance: *Frigoribacterium*, *Rhizobacter*, *Bifidobacterium*, *Leifsonia*, *Oerskovia*, *Xanthomonas*, *Streptomyces*, *Microbacterium*, *Turicella* and *Rhodopseudomonas*. These genera comprised around 50% of the total genera in each treatment (Supplementary Figure S3) and their relative abundance was significantly different for all (Figure 5).

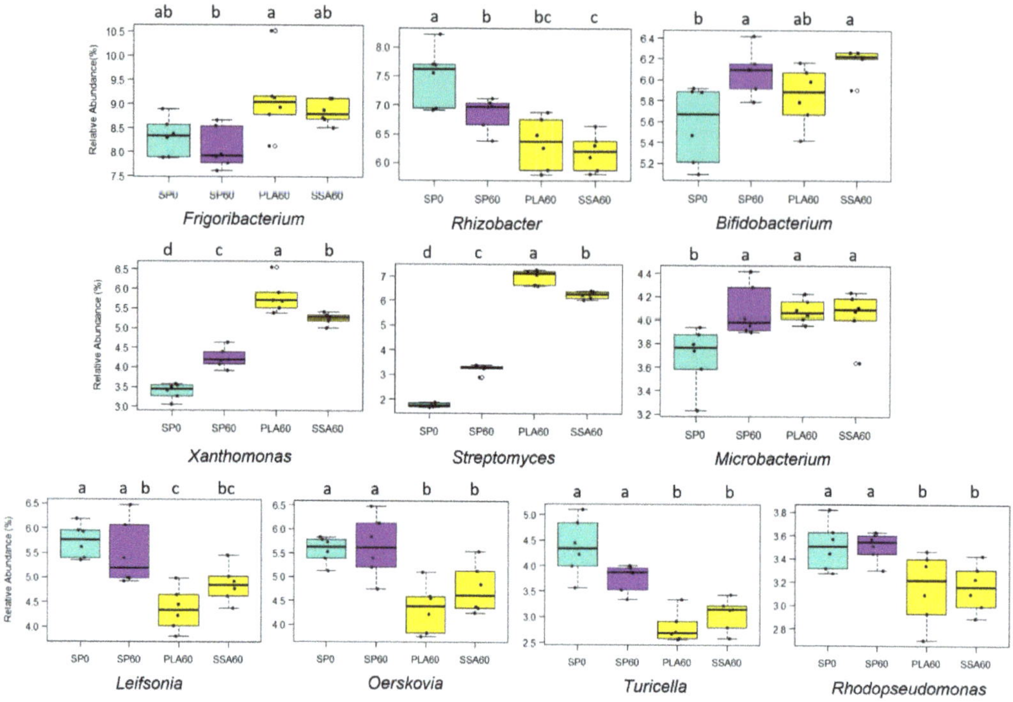

Figure 5. Mean relative abundance (%) bar plots of top 10 genera of the *phoD* functional amplicon sequencing with significantly different abundance, significance determined via Kruskal-Wallis test and Wilcoxon post-hoc analysis and Benjamini-Hochberg correction, different letters a–d state significant difference, P0 control (SP0) data depicted in light blue, mineral fertilizer (SP60) in purple and ash fertilizer treatments (PLA60, SSA60) in yellow, $n = 6$.

The genera *Leifsonia*, *Oerskovia*, *Rhodopseudomonas* and *Turicella* were significantly relatively lower in abundance ($p \leq 0.05$) in the ash RDF treatments compared to the control SP0 and SP60 fertilizer. *Rhizobacter* displayed the highest relative abundance in the P-free SP0 treatment. For the genera *Frigoribacterium*, *Streptomyces* and *Xanthomonas* the PLA60 treatment had the highest relative abundance, followed by SSA60, while the SP60 treatments displayed significantly lower relative abundances. The genus *Microbacterium* seemed to be positively affected by the P addition, all P-fertilized treatments had a significantly higher relative abundance compared to SP0. Lastly, the genus *Bifidobacterium* was significantly relative higher in abundance in the SP60 and SSA60 treatment, while it was similarly relative abundant in PLA60 and SP0.

4. Discussion

This study investigated the effects of ash RDF treatments at P concentration of 60 kg P ha^{-1} on a second growing season of perennial ryegrass alongside a positive (orthophosphate) and negative control (no P). Microbial capacities in rhizosphere P cycling were investigated that included phosphonate and phytate utilization, phosphatase activities and the related absolute gene copy numbers, as well as bacterial community structures were assessed (16S, *phoD*) alongside mycorrhization of ryegrass roots. The present hypothesis was that ash RDFs derived from incineration of poultry litter and sewage sludge positively affect the abundance and function of P mobilizing bacteria and arbuscular mycorrhizal fungi and have less impact on the soil microbial community than mineral P fertilizer in the long-term, while demonstrating plant growth promotion in *L. perenne* in a second growing season.

Sewage sludge and manure incineration is practiced as a sanitary way to destroy organic pollutants (pathogens, pharmaceutic compounds such as antibiotics, tranquilizers, diuretics, and hormones). In countries such as Switzerland and the Netherlands, 100% of sewage sludge is incinerated [49]. The resulting ash is nutrient-rich and under the right processing conditions it can be applied as fertilizer. The utilization of ashes derived from various nutrient-rich waste streams as fertilizer can be seen as product valorisation and likely reduces the amount of sewage sludge going to landfill, thus supporting the circular economy [50]. There is also potential for energy generation in the form of electricity or heat during the thermal process of such wastes, as described for the poultry litter ash used in this experiment [51]. Furthermore, the application of pyrolyzed biomass under oxygen limiting conditions (biochar), has shown positive effects in terms of soil fertility and the P cycling microbial community [52], raising interest in investigating the fate of ash application to soil. Due to the high incineration temperatures of 900 and 1000 °C respectively for the sewage (SSA) and poultry litter ash (PLA), only limited amounts of carbon is expected to remain in the ashes [24]. Instead, high levels of Ca, P, K, Mg, Fe and Al were determined in both ashes with the sewage sludge ash having particularly high levels of Fe and Al (Supplementary Information S1). Therefore, one may assume that most of the P in the ashes is present as Ca, Al or Fe -phosphate secondary mineral which may become plant available due to microbial dissolution activities [25]. The short-term study of the ashes fertilization that occurred prior to the present study investigated the mobilization of P from tri-calcium phosphates (TCP) as indicator for the microbial capacity to mobilize inorganically bound P [25]. There, significantly higher bacterial abundances were found in the PLA but not in the SSA treatment in comparison to the P-free control. However, both ashes were also applied in a field trial and there, TCP effects were only observed within the first three months of application, while from five months onwards, significant differences on TCP plates were no longer present [53]. The present microcosm study that reused fertilized soils to simulate a long-term ashes fertilizer event focused on the microbial processes of organically bound P mobilization (i.e., mineralization) under the assumption that significant proportions of the inorganic fertilizer P has been converted into organic forms by the time the microcosm study took place. Indeed in a previous grassland column study, phosphatase activity increased with orthophosphate fertilization, suggesting that significant amounts of inorganic P was first immobilized into organic forms [18] and then putatively mineralized again.

Treatment with mineral P (SP60) in the present study resulted in the lowest agronomic efficiency (difference method by yield) when compared to the ash treatments. However, AE does not account for the P plant uptake, which was highest in SP60. Statistical significance could not be determined for the plant elemental analysis, because the treatments had to be pooled to acquire a reasonable sample size for elemental analysis, thus further studies producing larger amounts of plant biomass would be needed to assess these trends. As already indicated above, both ashes from the present study were also investigated for their fertilizing effects at application rates of 40 kg P ha^{-1} in comparison to superphosphate and struvites under *L. perenne* cultivation in a field trial. There, no differences between treatments were detected regarding biomass yields within two years but apparent recovery efficiency (RE) of PLA was lower than for superphosphate fertilization (Patrick Forrestal, oral communication). While the present study is based on soils with low P availability (Irish P index of 2, which is typical for Ireland) and a single application of P fertilizer, it is worth noting that in other regions, soils may have a high P availability due to repeated fertilization. In the latter case, a P saturation may occur, where a larger fraction of the applied P will remain plant available [54]. According to Barrow [54] this higher plant availability is rooted in a saturation effect where inorganic P is sorbed onto and into soil particles to a point where sorption processes slow down and desorption increases to create a higher equilibrium of orthophosphate in soil solution. In the present study however, one can expect that both

PLA and SSA contain P in the form of particulate secondary minerals with Ca, Fe and Al. While the present study has not investigated the inorganic chemistry of soil and fertilizer P, we speculated that the presence of PLA and SSA in the soil beyond a growing season would lead to the (partial) transformation of the phosphate secondary minerals in the ashes into organic forms. Hence, this study investigated primarily (micro)biological factors in P cycling (i.e., mineralization).

In the present study, the P fertilization effect in a second growing season had significantly shifted the bacterial community in both the fingerprinting as well as the amplicon-based sequencing analysis for all treatments. The vesicular AMF colonization of the roots (VC) significantly affected the bacterial community structures as demonstrated by the DGGE and the amplicon-based sequencing of 16S rRNA gene fragments. Interestingly, at DGGE level VC appeared to separate SSA60 from the remaining bacterial communities, while for the amplicon-based analysis this was the case for PLA60. The remaining factors such as plant available P, dry matter yield, phosphonate and phytate utilization, *phoC* and *phoD* gene copy numbers and soil pH appeared to be important for the distinction of SP60 from the bacterial communities from the RDF treatments.

High supply rates of P for plant growth have been previously linked to reduced rates of root mycorrhization [55,56]. Even a single application of 20 kg P ha^{-1} to grassland columns significantly reduced arbuscular colonization (AC) [18]. In the present study SSA60 had the lowest P availability among the P fertilizer treatments while at the same time the highest levels of AC, although this was only a trend (did not reach significance). The opposite trend was observed for vesicular colonization (VC) in SSA60, which was the lowest in this study of all treatments. Vesicles function as lipid storage bodies and are described as a potential C sink [57]. Previously, limited photosynthate supply from the plant host has been identified to lower VC rates [58], hence this could indicate that SAA60 treated plants are nutrient limited, even with higher AC rates possibly due to low P availability. Further studies would be necessary to confirm this hypothesis as the current results are only trends and did not reach significance.

Soil pH had one of the strongest effects on the soil bacterial community. This is likely due to the retained liming effect of the ash RDF treatments. Even a year after their application, soil pH was still significantly higher than in SP60 and SP0, although the soil pH still had to be considered rather acidic (pH 5.3). Soil pH is acknowledged as one of the main stimulators of change for the microbial community structure [59] and it also affects P availability in soil [60] as well as plant growth [61]. Liming is a common practice in acidic temperate soils to improve P plant availability [62].

Neither ACP nor ALP activity in soil were affected by the treatments in the long term in this current study, but nonetheless, significant higher *phoC* copy numbers were detected in SSA60, and *phoD* copy numbers for both the PLA60 and SSA60 treatment. A negative correlation between ALP activity and P plant availability was described recently [63], while in the current study a significantly higher amount of plant available P for all P fertilized treatments did not result in a lower ALP activity. However, P mineralization processes can be steered not only through available P deficiencies in soils but by a demand for C instead [64]. External carbon inputs (manure, sewage sludge, cellulose) to soil have reportedly increased soil respiration and therefore indicate facilitated microbial growth [65]. Thus, the ash C input in the present study, in particular PLA60, could have contributed to the increased abundance of phosphonate and phosphate-ester utilizing bacteria.

Detailed analysis of the bacterial 16S rRNA gene analysis revealed two out of the ten highest relative abundant phyla were significantly higher in relative abundance in both ash treatments compared to SP0 and SP60, namely *Chloroflexi* and *Planctomycetes*. Additionally, PLA60 had significantly higher abundance of *Acidobacteria* and *Thaumarchaeota* compared to SP0 and SP60. The phyla *Acidobacteria*, *Planctomycetes*, and *Chloroflexi* were also described as part of the *phoX* (phosphatase)-harboring community [66]. Likewise, *Actinobacteria*, *Firmicutes* and *Planctomycetes*, that were some of the highest relative abundant phyla in this current study, have also been found to commonly contain *phoD* [66]. However, *Firmicutes* were significantly lower in relative abundance in the ash RDF treatments, while *Actinobacteria* relative abundance was not significantly affected by any of the treatments. *Actinobacteria* were reported to be effective inorganic P solubilizer via organic acid exudation [67]. Increasing amounts of mineral P applications (5, 10, 20 kg P ha^{-1}) have previously reduced relative abundance of *Actinobacteria* and *Firmicutes* in grassland columns [18], which suggests that mineral P application can negatively affect both phyla.

To study the functional diversity at the phosphatase gene level, the present study decided to focus on *phoD* over *phoC* as for *phoD*, both ash treatments resulted in significantly higher gene abundances over the control and conventional P fertilizer. Furthermore, the same type of analysis was carried out with the same soils previously [25] as a short-term study, where no significant shifts of the *phoD* community were identified. This was different to the present study, where the presumptive long-term effect affected the *phoD* community structure. Significant separation of the *phoD*-harboring community in this current study was driven by the same environmental parameters as the overall bacterial structure, except for VC. In this current study, ash fertilization demonstrated positive as well as negative effects on the relative abundance of *phoD*-harboring genera. Some of the genera with higher relative abundance for the ash treatments were generally also predominantly abundant in other studies investigating organic fertilization. The genus *Xanthomonas* was also detected in chicken manure and vegetable waste composting samples [68]. *Xanthomonas phoD* sequences were also identified in black truffle inoculated pine seedlings [69]. *Streptomyces* was one of the genera dominating long-term organically fertilized soil in the Jiangxi Province in China [70] and was one of the abundant *phoD* harbouring bacteria in composts containing biochar [71]. Yet to date, neither genus has been reported in perennial ryegrass in connection to *phoD*. It therefore transpires that both genera may represent important members of the phosphate-ester mineralizing communities that are related to fertilizers and/or soil improvers. Therefore, their presence may suggest a potential benefit for plant P availability.

5. Conclusions

In conclusion, ash RDF application was beneficial to the abundance of microbial P mobilizers, as indicated by significantly higher PAA and phytate utilization and higher *phoC* and *phoD* copy numbers and a shift in the *phoD* harbouring community, although this did not translate into a higher phosphatase activity in soil compared to the mineral P treatment. However, Pav was also increased in the ash treatments, likely due to P release from phosphonates and phytates and high ACP activity in general, therefore it can be assumed that the *phoD*-harbouring community in the ash RDF treatments might induce ALP production when the soil becomes more P deficient. At the same time, the mineral SP60 treatment demonstrated reduced microbial P mobilization capabilities, which suggests a higher dependency on external P inputs upon plant available P depletion. Further studies at field scale are necessary to study the potential use of ashes as P fertilizer. A particularly interesting option would be to test combinations of quickly available P sources with ashes in order to see whether the putative beneficial effects of the ashes at the microbial level can be maintained while at the same time optimising crop yields.

Supplementary Materials: The following supporting information can be downloaded at: https://www.mdpi.com/article/10.3390/environments11030049/s1, Figure S1. Example of arbuscules (A) and vesicles (B) in *L. perenne* roots after cultivation in soils fertilized with ashes, visualized at 600× magnification (BX60, Olympus, Tokyo, Japan); Figure S2: Mean relative abundance of the top 10 abundant genera obtained from 16S rRNA amplicon sequencing of the microcosm trial, all other genera collapsed in "others", $n = 6$; Figure S3: Mean relative abundance of the top 10 abundant genera obtained from *phoD* amplicon sequencing of the microcosm trial, all other genera collapsed in "others", $n = 6$. Table S1: AMF colonization of the *L. perenne* roots after cultivation in microcosms for 54 days in soil amended with P fertilizers SP, PLA and SSA a year prior to the experiment, arbuscular (AC), vesicular (VC) and hyphal (HC) colonization were recorded via microscopic investigation at 600× magnification after root staining with trypan blue, 100 observations per sample, ± represents standard error, n = 6; Table S2: Alpha diversity measures of 16S rRNA amplicon sequencing results of the microcosm trial, significance determined via Kruskal-Wallis test and Wilcoxon post-hoc analysis and Benjamini–Hochberg correction, different letters within a column indicate statistically significant difference, n = 6; Table S3: Alpha diversity measures of *phoD* functional gene sequencing results of the microcosm trial, significance determined via Kruskal–Wallis test and Wilcoxon post-hoc analysis and Benjamini-Hochberg correction, different letters within a column indicate statistically significant difference, n = 6. Reference [72] is cited in the supplementary materials.

Author Contributions: Conceptualization, A.S.; methodology, A.S. and L.D.; software, L.D.; validation, L.D.; formal analysis, L.D.; investigation, L.D.; resources, A.S.; data curation, L.D. and A.S.; writing—original draft preparation, L.D.; writing—review and editing, A.S. and L.D.; visualization, L.D.; supervision, A.S.; project administration, A.S.; funding acquisition, A.S. All authors have read and agreed to the published version of the manuscript.

Funding: This research was funded by the North-West Europe Interreg project ReNu2Farm Nutrient Recycling—from pilot production to farms and fields (NWE601).

Data Availability Statement: All sequencing data have been deposited in the Sequence Read Archive (SRA) and is available in BioProject PRJNA1043576 (SUB13990651).

Acknowledgments: We would like to acknowledge the ReNu2Farm consortium members for their general support in completing this project as well as the technical team of Biological Sciences.

Conflicts of Interest: The authors declare no conflicts of interest. The funders had no role in the design of the study; in the collection, analyses, or interpretation of data; in the writing of the manuscript; or in the decision to publish the results.

References

1. United Nations. Growing at a Slower Pace, World Population is Expected to Reach 9.7 Billion in 2050 and Could Peak at Nearly 11 Billion around 2100. Available online: https://www.un.org/sustainabledevelopment/blog/2019/06/growing-at-a-slower-pace-world-population-is-expected-to-reach-9-7-billion-in-2050-and-could-peak-at-nearly-11-billion-around-2100-un-report/ (accessed on 24 January 2022).
2. Liu, Y.; Villalba, G.; Ayres, R.U.; Schroder, H. Global Phosphorus Flows and Environmental Impacts from a Consumption Perspective. *J. Ind. Ecol.* **2008**, *12*, 229–247. [CrossRef]
3. Reijnders, L. Phosphorus Resources, Their Depletion and Conservation, A Review. *Resour. Conserv. Recycl.* **2014**, *93*, 32–49 [CrossRef]
4. Ruttenberg, K.C. The Global Phosphorus Cycle. In *Treatise on Geochemistry*; Elsevier: Amsterdam, The Netherlands, 2003, pp. 585–643.
5. Smit, A.L.; Bindraban, P.S.; Schröder, J.J.; Conijn, J.G.; van der Meer, H.G. *Phosphorus in Agriculture: Global Resources, Trends and Developments*; Plant Research International: Wageningen, The Netherlands, 2009.
6. Geissdoerfer, M.; Savaget, P.; Bocken, N.M.P.; Hultink, E.J. The Circular Economy—A New Sustainability Paradigm? *J. Clean. Prod.* **2017**, *143*, 757–768. [CrossRef]
7. Powers, S.M.; Chowdhury, R.B.; MacDonald, G.K.; Metson, G.S.; Beusen, A.H.W.; Bouwman, A.F.; Hampton, S.E.; Mayer, B.K.; McCrackin, M.L.; Vaccari, D.A. Global Opportunities to Increase Agricultural Independence Through Phosphorus Recycling. *Earth's Future* **2019**, *7*, 370–383. [CrossRef]
8. Mackey, K.; Paytan, A. Phosphorus Cycle. *Encycl. Microbiol.* **2009**, *3*, 322–334.
9. Barrow, N.J. Comparing two theories about the nature of soil phosphate. *Eur. J. Soil Sci.* **2021**, *72*, 679–685. [CrossRef]
10. Gerke, J. Phytate (Inositol Hexakisphosphate) in Soil and Phosphate Acquisition from Inositol Phosphates by Higher Plants. A Review. *Plants* **2015**, *4*, 253–266. [CrossRef] [PubMed]
11. Barea, J.-M.; Richardson, A.E. Phosphate Mobilisation by Soil Microorganisms. In *Principles of Plant-Microbe Interactions*; Lugtenberg, B., Ed.; Springer International Publishing: Cham, Switzerland, 2015; pp. 225–234.
12. Bindraban, P.S.; Dimkpa, C.O.; Pandey, R. Exploring Phosphorus Fertilizers and Fertilization Strategies for Improved Human and Environmental Health. *Biol. Fertil. Soils* **2020**, *56*, 299–317. [CrossRef]
13. Wang, T.; Camps-Arbestain, M.; Hedley, M.; Bishop, P. Predicting Phosphorus Bioavailability from High-Ash Biochars. *Plant Soil* **2012**, *357*, 173–187. [CrossRef]
14. Richardson, A.E.; Simpson, R.J. Soil Microorganisms Mediating Phosphorus Availability Update on Microbial Phosphorus. *Plant Physiol.* **2011**, *156*, 989–996. [CrossRef] [PubMed]
15. Seeling, B.; Zasoski, R.J. Microbial Effects in Maintaining Organic and Inorganic Solution Phosphorus Concentrations in a Grassland Topsoil. *Plant Soil* **1993**, *148*, 277–284. [CrossRef]
16. Mozumder, P.; Berrens, R.P. Inorganic Fertilizer Use and Biodiversity Risk: An Empirical Investigation. *Ecol. Econ.* **2007**, *62*, 538–543. [CrossRef]

17. Liu, J.; Zhang, J.; Li, D.; Xu, C.; Xiang, X. Differential Responses of Arbuscular Mycorrhizal Fungal Communities to Mineral and Organic Fertilization. *MicrobiologyOpen* **2020**, *9*, e00920. [CrossRef] [PubMed]
18. Ikoyi, I.; Fowler, A.; Schmalenberger, A. One-Time Phosphate Fertilizer Application to Grassland Columns Modifies the Soil Microbiota and Limits Its Role in Ecosystem Services. *Sci. Total Environ.* **2018**, *630*, 849–858. [CrossRef] [PubMed]
19. Huotari, N.; Tillman-Sutela, E.; Moilanen, M.; Laiho, R. Recycling of Ash—For the Good of the Environment? *For. Ecol. Manag.* **2015**, *348*, 226–240. [CrossRef]
20. Rosenberg, O.; Persson, T.; Högbom, L.; Jacobson, S. Effects of Wood-Ash Application on Potential Carbon and Nitrogen Mineralisation at Two Forest Sites with Different Tree Species, Climate and N Status. *For. Ecol. Manag.* **2010**, *260*, 511–518. [CrossRef]
21. Jacobson, S.; Lundström, H.; Nordlund, S.; Sikström, U.; Pettersson, F. Is Tree Growth in Boreal Coniferous Stands on Mineral Soils Affected by the Addition of Wood Ash? *Scand. J. For. Res.* **2014**, *29*, 675–685. [CrossRef]
22. Perkiömäki, J.; Fritze, H. Does Simulated Acid Rain Increase the Leaching of Cadmium from Wood Ash to Toxic Levels to Coniferous Forest Humus Microbes? *FEMS Microbiol. Ecol.* **2003**, *44*, 27–33. [CrossRef] [PubMed]
23. Huygens, D.; Saveyn, H.G.M. Agronomic Efficiency of Selected Phosphorus Fertilisers Derived from Secondary Raw Materials for European Agriculture. A Meta-Analysis. *Agron. Sustain. Dev.* **2018**, *38*, 52. [CrossRef]
24. Hermann, L.; Schaaf, T. Outotec (AshDec®) Process for P Fertilizers from Sludge Ash. In *Phosphorus Recovery and Recycling*; Ohtake, H., Tsuneda, S., Eds.; Springer: Singapore, 2019; pp. 221–233.
25. Deinert, L.; Ikoyi, I.; Egeter, B.; Forrestal, P.; Schmalenberger, A. Short-Term Impact of Recycling-Derived Fertilizers on Their P Supply for Perennial Ryegrass (*Lolium perenne*). *Plants* **2023**, *12*, 2762. [CrossRef]
26. Gahan, J.; O'Sullivan, O.; Cotter, P.D.; Schmalenberger, A. Arbuscular Mycorrhiza Support Plant Sulfur Supply through Organosulfur Mobilizing Bacteria in the Hypho- and Rhizosphere. *Plants* **2022**, *11*, 3050. [CrossRef]
27. Murphy, J.; Riley, J.P. A Modified Single Solution Method for the Determination of Phosphate in Natural Waters. *Anal. Chim. Acta* **1962**, *27*, 31–36. [CrossRef]
28. Tabatabai, M.A.; Bremner, J.M. Use of p-Nitrophenyl Phosphate for Assay of Soil Phosphatase Activity. *Soil Biol. Biochem.* **1969**, *1*, 301–307. [CrossRef]
29. Fox, A.; Kwapinski, W.; Griffiths, B.S.; Schmalenberger, A. The Role of Sulfur- and Phosphorus-Mobilizing Bacteria in Biochar-Induced Growth Promotion of *Lolium perenne*. *FEMS Microbiol. Ecol.* **2014**, *90*, 78–91. [CrossRef] [PubMed]
30. McGonigle, T.P.; Miller, M.H.; Evans, D.G.; Fairchild, G.L.; Swan, J.A. A New Method Which Gives an Objective Measure of Colonization of Roots by Vesicular-Arbuscular Mycorrhizal Fungi. *New Phytol.* **1990**, *115*, 495–501. [CrossRef] [PubMed]
31. Muyzer, G.; de Waal, E.C.; Uitterlinden, A.G. Profiling of Complex Microbial Populations by Denaturing Gradient Gel Electrophoresis Analysis of Polymerase Chain Reaction-Amplified Genes Coding for 16S rRNA. *Appl. Environ. Microbiol.* **1993**, *59*, 695–700. [CrossRef] [PubMed]
32. Ragot, S.A.; Kertesz, M.A.; Bünemann, E.K. *phoD* Alkaline Phosphatase Gene Diversity in Soil. *Appl. Environ. Microbiol.* **2015**, *81*, 7281–7289. [CrossRef] [PubMed]
33. Fraser, T.D.; Lynch, D.H.; Gaiero, J.; Khosla, K.; Dunfield, K.E. Quantification of Bacterial Non-Specific Acid (*phoC*) and Alkaline (*phoD*) Phosphatase Genes in Bulk and Rhizosphere Soil from Organically Managed Soybean Fields. *Appl. Soil Ecol.* **2017**, *111*, 48–56. [CrossRef]
34. Sakurai, M.; Wasaki, J.; Tomizawa, Y.; Shinano, T.; Osaki, M. Analysis of Bacterial Communities on Alkaline Phosphatase Genes in Soil Supplied with Organic Matter. *Soil Sci. Plant Nutr.* **2008**, *54*, 62–71. [CrossRef]
35. Bolyen, E.; Rideout, J.R.; Dillon, M.R.; Bokulich, N.A.; Abnet, C.C.; Al-Ghalith, G.A.; Alexander, H.; Alm, E.J.; Arumugam, M.; Asnicar, F.; et al. Reproducible, Interactive, Scalable and Extensible Microbiome Data Science Using QIIME 2. *Nat. Biotechnol.* **2019**, *37*, 852–857. [CrossRef] [PubMed]
36. Callahan, B.J.; McMurdie, P.J.; Rosen, M.J.; Han, A.W.; Johnson, A.J.A.; Holmes, S.P. DADA2: High-Resolution Sample Inference from Illumina Amplicon Data. *Nat. Methods* **2016**, *13*, 581–583. [CrossRef] [PubMed]
37. Katoh, K.; Standley, D.M. MAFFT Multiple Sequence Alignment Software Version 7: Improvements in Performance and Usability. *Mol. Biol. Evol.* **2013**, *30*, 772–780. [CrossRef] [PubMed]
38. Price, M.N.; Dehal, P.S.; Arkin, A.P. FastTree 2—Approximately Maximum-Likelihood Trees for Large Alignments. *PLoS ONE* **2010**, *5*, e9490. [CrossRef] [PubMed]
39. Faith, D.P. Conservation Evaluation and Phylogenetic Diversity. *Biol. Conserv.* **1992**, *61*, 1–10. [CrossRef]
40. Lozupone, C.A.; Hamady, M.; Kelley, S.T.; Knight, R. Quantitative and Qualitative β Diversity Measures Lead to Different Insights into Factors That Structure Microbial Communities. *Appl. Environ. Microbiol.* **2007**, *73*, 1576–1585. [CrossRef] [PubMed]
41. Bokulich, N.A.; Kaehler, B.D.; Rideout, J.R.; Dillon, M.; Bolyen, E.; Knight, R.; Huttley, G.A.; Gregory Caporaso, J. Optimizing Taxonomic Classification of Marker-Gene Amplicon Sequences with QIIME 2's q2-feature-classifier Plugin. *Microbiome* **2018**, *6*, 90. [CrossRef] [PubMed]
42. Quast, C.; Pruesse, E.; Yilmaz, P.; Gerken, J.; Schweer, T.; Yarza, P.; Peplies, J.; Glöckner, F.O. The SILVA ribosomal RNA Gene Database Project: Improved Data Processing and Web-Based Tools. *Nucleic Acids Res.* **2013**, *41*, D590–D596. [CrossRef] [PubMed]
43. Yilmaz, P.; Parfrey, L.W.; Yarza, P.; Gerken, J.; Pruesse, E.; Quast, C.; Schweer, T.; Peplies, J.; Ludwig, W.; Glöckner, F.O. The SILVA and "All-species Living Tree Project (LTP)" taxonomic frameworks. *Nucleic Acids Res.* **2014**, *42*, D643–D648. [CrossRef] [PubMed]
44. Martin, M. Cutadapt Removes Adapter Sequences from High-Throughput Sequencing Reads. *EMBnet J.* **2011**, *17*, 10. [CrossRef]

45. Edgar, R.C. Search and Clustering Orders of Magnitude Faster than BLAST. *Bioinformatics* **2010**, *26*, 2460–2461. [CrossRef] [PubMed]
46. Edgar, R.C. UPARSE: Highly Accurate OTU Sequences from Microbial Amplicon Reads. *Nat. Methods* **2013**, *10*, 996–998. [CrossRef] [PubMed]
47. Fraser, T.D.; Lynch, D.H.; Bent, E.; Entz, M.H.; Dunfield, K.E. Soil Bacterial *phoD* Gene Abundance and Expression in Response to Applied Phosphorus and Long-Term Management. *Soil Biol. Biochem.* **2015**, *88*, 137–147. [CrossRef]
48. Fish, J.A.; Chai, B.; Wang, Q.; Sun, Y.; Brown, C.T.; Tiedje, J.M.; Cole, J.R. FunGene: The Functional Gene Pipeline and Repository. *Front. Microbiol.* **2013**, *4*, 291. [CrossRef] [PubMed]
49. Herzel, H.; Krüger, O.; Hermann, L.; Adam, C. Sewage Sludge Ash—A Promising Secondary Phosphorus Source for Fertilizer Production. *Sci. Total Environ.* **2016**, *542*, 1136–1143. [CrossRef] [PubMed]
50. Gherghel, A.; Teodosiu, C.; De Gisi, S. A Review on Wastewater Sludge Valorisation and Its Challenges in the Context of Circular Economy. *J. Clean. Prod.* **2019**, *228*, 244–263. [CrossRef]
51. Choudhury, A.; Felton, G.; Moyle, J.; Lansing, S. Fluidized bed combustion of poultry litter at farm-scale: Environmental impacts using a life cycle approach. *J. Clean. Prod.* **2020**, *276*, 124231. [CrossRef]
52. Lehmann, J.; Rillig, M.C.; Thies, J.; Masiello, C.A.; Hockaday, W.C.; Crowley, D. Biochar Effects on Soil Biota—A Review. *Soil Biol. Biochem.* **2011**, *43*, 1812–1836. [CrossRef]
53. Deinert, L.; Ashekuzzaman, S.; Forrestal, P.; Schmalenberger, A. The impact of struvite and ash recycling-derived fertilizers on microbial phosphorus mobilization capabilities and community structure in a *Lolium perenne* field trial. *bioRxiv* **2024**, arXiv:bioRxiv:2024.2001.2016.575387.
54. Barrow, N.J. How understanding soil chemistry can lead to better phosphate fertilizer practice: A 68 year journey (so far). *Plant Soil* **2022**, *476*, 117–131. [CrossRef]
55. Bakhshandeh, S.; Corneo, P.E.; Mariotte, P.; Kertesz, M.A.; Dijkstra, F.A. Effect of Crop Rotation on Mycorrhizal Colonization and Wheat Yield Under Different Fertilizer Treatments. *Agric. Ecosyst. Environ.* **2017**, *247*, 130–136. [CrossRef]
56. Gryndler, M.; Larsen, J.; Hršelová, H.; Řezáčová, V.; Gryndlerová, H.; Kubát, J. Organic and Mineral Fertilization, Respectively, Increase and Decrease the Development of External Mycelium of Arbuscular Mycorrhizal Fungi in a Long-Term Field Experiment. *Mycorrhiza* **2006**, *16*, 159–166. [CrossRef] [PubMed]
57. Dodd, J.C. Mycelium of Arbuscular Mycorrhizal Fungi (AMF) From Different Genera: Form, Function and Detection. *Plant Soil* **2000**, *226*, 131–151. [CrossRef]
58. Thompson, J.P. Soilless Culture of Vesicular–Arbuscular Mycorrhizae of Cereals: Effects of Nutrient Concentration and Nitrogen Source. *Can. J. Bot.* **1986**, *64*, 2282–2294. [CrossRef]
59. Tripathi, B.M.; Stegen, J.C.; Kim, M.; Dong, K.; Adams, J.M.; Lee, Y.K. Soil pH Mediates the Balance Between Stochastic and Deterministic Assembly of Bacteria. *ISME J.* **2018**, *12*, 1072–1083. [CrossRef] [PubMed]
60. Penn, C.J.; Camberato, J.J. A Critical Review on Soil Chemical Processes that Control How Soil pH Affects Phosphorus Availability to Plants. *Agriculture* **2019**, *9*, 120. [CrossRef]
61. Barrow, N.J.; Lambers, H. Phosphate-solubilising microorganisms mainly increase plant phosphate uptake by effects of pH on root physiology. *Plant Soil* **2022**, *476*, 397–402. [CrossRef]
62. Mijangos, I.; Albizu, I.; Epelde, L.; Amezaga, I.; Mendarte, S.; Garbisu, C. Effects of Liming on Soil Properties and Plant Performance of Temperate Mountainous Grasslands. *J. Environ. Manag.* **2010**, *91*, 2066–2074. [CrossRef]
63. Fraser, T.; Lynch, D.H.; Entz, M.H.; Dunfield, K.E. Linking Alkaline Phosphatase Activity with Bacterial *phoD* Gene Abundance in Soil from a Long-Term Management Trial. *Geoderma* **2015**, *257–258*, 115–122. [CrossRef]
64. Spohn, M.; Treichel, N.S.; Cormann, M.; Schloter, M.; Fischer, D. Distribution of Phosphatase Activity and Various Bacterial Phyla in the Rhizosphere of *Hordeum vulgare* L. Depending on P Availability. *Soil Biol. Biochem.* **2015**, *89*, 44–51. [CrossRef]
65. Gómez-Muñoz, B.; Efthymiou, A.; Dubey, M.; Sølve, J.; Nicolaisen, M.; Jensen, D.F.; Nybroe, O.; Larsen, J. Cellulose Amendment Promotes P Solubilization by *Penicillium aculeatum* in Non-Sterilized Soil. *Fungal Biol.* **2022**, *126*, 356–365. [CrossRef] [PubMed]
66. Ragot, S.A.; Kertesz, M.A.; Mészáros, É.; Frossard, E.; Bünemann, E.K. Soil *phoD* and *phoX* Alkaline Phosphatase Gene Diversity Responds to Multiple Environmental Factors. *FEMS Microbiol. Ecol.* **2017**, *93*, fiw212. [CrossRef] [PubMed]
67. Solans, M.; Messuti, M.I.; Reiner, G.; Boenel, M.; Vobis, G.; Wall, L.G.; Scervino, J.M. Exploring the Response of Actinobacteria to the Presence of Phosphorus Salts Sources: Metabolic and Co-Metabolic Processes. *J. Basic Microbiol.* **2019**, *59*, 487–495. [CrossRef] [PubMed]
68. Wan, W.; Wang, Y.; Tan, J.; Qin, Y.; Zuo, W.; Wu, H.; He, H.; He, D. Alkaline Phosphatase-Harboring Bacterial Community and Multiple Enzyme Activity Contribute to Phosphorus Transformation During Vegetable Waste and Chicken Manure Composting. *Bioresour. Technol.* **2020**, *297*, 122406. [CrossRef] [PubMed]
69. Zhang, X.; Li, X.; Ye, L.; Huang, Y.; Kang, Z.; Zhang, B.; Zhang, X. Colonization by *Tuber melanosporum* and *Tuber indicum* affects the growth of *Pinus armandii* and *phoD* alkaline phosphatase encoding bacterial community in the rhizosphere. *Microbiol. Res.* **2020**, *239*, 126520. [CrossRef]
70. Liu, W.; Ling, N.; Luo, G.; Guo, J.; Zhu, C.; Xu, Q.; Liu, M.; Shen, Q.; Guo, S. Active *phoD*-Harboring Bacteria are Enriched by Long-Term Organic Fertilization. *Soil Biol. Biochem.* **2021**, *152*, 108071. [CrossRef]

71. Yin, Y.; Yang, C.; Li, M.; Yang, S.; Tao, X.; Zheng, Y.; Wang, X.; Chen, R. Biochar reduces bioavailability of phosphorus during swine manure composting: Roles of *phoD*-harboring bacterial community. *Sci. Total Environ.* **2023**, *858*, 159926. [CrossRef]
72. Sigurnjak, I.; Saju, A.; Schmalenberger, A.; Lagrange, H.; Forrestal, P.; Postma, R.; van Scholl, L.; Verleden, I.; Ryan, D.; Krapinska, A.; et al. Product Information Sheets (WPT1_D3.4). Interreg North-West Europe; ReNu2Farm. Available online: https://www.nmi-agro.nl/wp-content/uploads/2019/09/21022022_WPT1_D3.4-Product-factsheets.pdf (accessed on 26 January 2024).

environments

MDPI

Article

The Effects of Different Combinations of Cattle Organic Soil Amendments and Copper on Lettuce (cv. *Rufus*) Plant Growth

Chiara De Carolis [1,2,3], Valentina Iori [1], Alessandra Narciso [2], Davide Gentile [1,4], Barbara Casentini [2], Fabrizio Pietrini [5], Paola Grenni [2], Anna Barra Caracciolo [2] and M. Adelaide Iannelli [1,*]

[1] Institute of Agricultural Biology and Biotechnology—National Research Council (CNR-IBBA), SP 35d, 9, Montelibretti, 00010 Rome, Italy; chiara.decarolis@uniroma1.it (C.D.C.); valentina.iori@cnr.it (V.I.); davide.gentile@uniroma1.it (D.G.)

[2] Water Research Institute, National Research Council (CNR-IRSA), SP 35d, 9, Montelibretti, 00010 Rome, Italy; alessandra.narciso@irsa.cnr.it (A.N.); barbara.casentini@irsa.cnr.it (B.C.); paola.grenni@irsa.cnr.it (P.G.); anna.barracaracciolo@irsa.cnr.it (A.B.C.)

[3] Department of Environmental Biology, 'Sapienza' University of Rome, P.le Aldo Moro 5, 00185 Rome, Italy

[4] Department of Biology and Biotechnology 'Charles Darwin', Sapienza University of Rome, P.le Aldo Moro 5, 00185 Rome, Italy

[5] Research Institute on Terrestrial Ecosystems—National Research Council (CNR-IRET), SP 35d, 9, Montelibretti, 00010 Rome, Italy; fabrizio.pietrini@cnr.it

* Correspondence: mariaadelaide.iannelli@cnr.it

Abstract: In modern agricultural production, cattle manure waste recovery is considered as a sustainable approach to agricultural waste management, reducing environmental pollution and chemical fertilizer use. This study aimed to investigate the effects of manure and digestate derived from a pilot-scale livestock waste-recycling system, in combination with a low copper concentration as a fungicide, on the physiological response of lettuce cv *Rufus* (*Lactuca sativa* L.) plants and the associated soil microbiome. A five-week microcosm experiment was conducted in a greenhouse under environmental conditions. Lettuce plant performance was assessed in terms of biomass, leaf area index, photosynthetic activity, chlorophyll measurements, lipid peroxidation, total phenolic content, and nutrient uptake. The results suggested that incorporating digestate into the potting soil mix significantly enhanced crop yields compared to the control and manure treatments. The soil microbial activity increased in the presence of fertilizers, improving the soil chemical and biological properties. The addition of copper negatively affected the growth and physiological performance of the lettuce plants under both the control and manure-treated conditions, except for those grown in the presence of digestate, where copper accumulation was reduced. These findings highlight the potential of growing horticultural crops using organic fertilization through livestock waste anaerobic digestate, establishing a waste-to-food recycling system.

Keywords: manure; digestate; copper; lettuce; plant growth; nutrient uptake

1. Introduction

As the human population increases, so does the demand for food, leading to a significant increase in food waste and its environmental impact. Over the last 50 years, worldwide, the use of inorganic fertilizers has significantly increased agricultural production. However, the use of organic fertilizers could be a sustainable way to reuse both organic wastes and manure from livestock farms in order to increase food production. Animal feces and urine contain high percentages of nitrogen (N) (55–95%), phosphorous (P), and potassium (K), which are essential macronutrients for animals, plants, and bacteria [1].

Although livestock waste is a valuable resource and can be used as a biofertilizer, it must be managed correctly to avoid environmental negative impacts, such as water and soil contamination (e.g., excess of organic nutrients and presence of trace micropollutants as antibiotics), air pollution (greenhouse gas emissions), and human health problems (spread of pathogens) [1]. A growing number of intensive livestock farms are turning to anaerobic digestion (AD) technology to solve the problem concerning the disposal of the huge volumes of manure and slurry produced every day. The produced biogas and digestate provide significant energy and fertilizer benefits [2]. Anaerobic digestion is an efficient process in which organic matter breaks down naturally in the absence of oxygen and takes place in a reactor, where the resulting greenhouse gases can be reused rather than released into the atmosphere [1]. The by-product of the AD process is digestate, which can be employed as a biofertilizer returning nutrients to the soil through natural cycles and replacing inorganic fertilizer inputs. Thus, AD can be considered as a process that reduces the use of fossil fuels and inorganic fertilizers, lowers greenhouse gas emissions, and provides a highly efficient method for recycling resources, in line with green and circular economy projects [1,3].

Previous studies have demonstrated that digestate can be a source of macro and micronutrients useful for improving soil quality, promoting biological activity, and improving physiological responses in crop growth [4–6]. Nevertheless, digestate, due to its chemical compounds, can also cause slow growth and early senescence in crop plants [7].

Copper (Cu) is an essential micronutrient for plant growth and metabolism, related to many physiological and biochemical processes in plants [8,9]. As a cofactor for many enzymes, Cu plays a key role under stress conditions. Copper is also associated with oxidative phosphorylation, protein trafficking, signal regulation, and lipid and iron metabolism [10]. Indeed, Cu deficiencies can alter several functions of plant metabolism and its excessive concentration can generate phytotoxicity, resulting in the overproduction of reactive oxygen species (ROS) and causing damage to carbohydrates, lipids, proteins, and DNA in respiration, photosynthesis, and enzyme activity that can result in growth and productivity inhibition and imbalances in nutrient uptake [11,12]. Actually, owing to different agricultural practices, such as the use of manure as fertilizer (especially of pig origin) and copper application as a pesticide in vineyards, this metal can accumulate in high concentrations in soils and can become a pollutant [8,9].

In particular, Cu has been applied to vineyards, orchards, and horticultural crops since the late 19th century at doses of several kilograms per hectare to prevent downy mildew [9]. Copper's antifungal properties were demonstrated in the laboratory in the 1950s, and nowadays, its use is also authorized in organic farming. Copper-based fungicides, such as the Bordeaux mixture, consist of copper (II) sulphate ($CuSO_4$) and quicklime (CaO). Indeed, the average Cu concentration in European agricultural soils has been increasing and it has been estimated to be between 30 and 50 mg kg^{-1} [8]. The use of copper in agricultural soils has been restricted by the European Commission (EU Reg. 1981/2018) to minimize its potential accumulation in the soil and exposure to non-target organisms, such as plants and animals, so a maximum of 28 kg ha^{-1} may be applied for plant protection over a 7-year period, with an annual limit of 4 kg ha^{-1} [9].

Possible copper mobility from soil to plant tissue raises several concerns, including possible adverse effects on herbivorous organisms and humans, which ingest edible plants [9,13].

Lettuce (*Lactuca sativa* L.), an annual crop belonging to the Asteraceae family, is among the world's most widely cultivated and consumed vegetables and it is rich in nutrients as well as vitamins, calcium, and fiber. It also has a short growing season, characteristics for which it is considered a model vegetable species [14,15]. Moreover, lettuce is a biological species recommended for soil toxicity testing by the OECD (1984) to evaluate the potential impact of using natural products to limit contaminants and plant growth [16].

The present study aimed to investigate the effects of the application of different combinations of cattle organic soil amendments, manure (M), or digestate (D), and low concentrations of copper on the growth of lettuce plants in terms of biometric (i.e., shoot biomass, shoot height, number of leaves, and surface leaf area), photosynthetic (i.e., chlorophyll content and fluorescence), and biochemical (i.e., lipid peroxidation and total phenol content) parameters, and the uptake of macro- and micro-nutrients highlighting the copper element. Furthermore, the response of the associated soil microbial community (i.e, soil total microbial abundance and soil microbial activity) was analyzed.

2. Materials and Methods

2.1. Plant, Soil Materials, and Experimental Design

Lettuce (*Lactuca sativa* L.), as one of the most economically important leafy vegetables, was chosen as the model plant species. Twenty-four seedlings of lettuce (cv. "Rufus"), with a height of about 4 cm, were purchased from a local hatchery. The seedlings were transplanted into plastic pots (1 seedling/pot) (1.4 L capacity), previously filled with 900 g of soil. Before transplanting, cattle manure (1% *w*/*w*) or digestate (1% *w*/*w*) was homogenously added to the soil alone or in combination with 30 mg kg^{-1} of copper sulphate ($CuSO_4$ $5H_2O$). All residual vegetable matter and stones were manually removed from the soil, and it was air-dried before setting up the microcosm experiment. Distilled water was used to obtain a 29.8% soil humidity, corresponding to 45% of the soil water-holding capacity. Soil was originally collected from an organic local farm, while manure and digestate were obtained from a biogas plant located on a cattle farm in an agricultural area in central Italy. The main physico-chemical characteristics of the soil, manure, and digestate, such as the pH, total organic carbon (Corg) and nitrogen (N) (Table S1), and elements content (Table S2), were measured according to the methods described in [17]. Afterwards, the seedlings were placed in a greenhouse located within the National Research Council (CNR) (42°10′ N, 12°63′ E) in Montelibretti (Rome, Italy), and grown for five weeks (from May to June 2022), where there was a daily average temperature of 26–31 °C during the day and 12–17 °C at night (Figure 1).

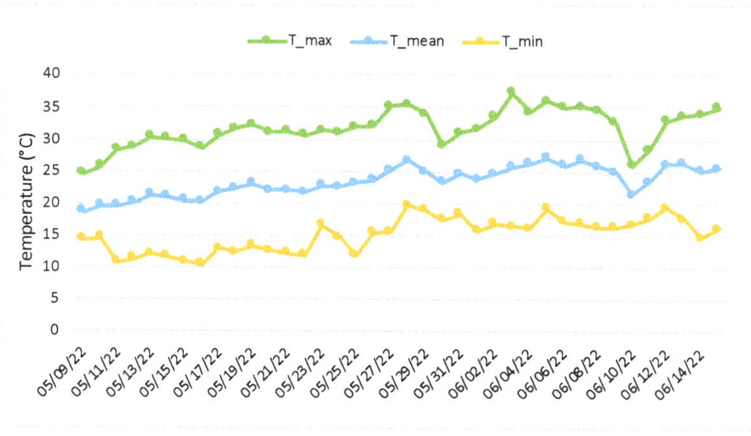

Figure 1. Temperatures observed during the experimental period in the greenhouse: Maximum (T_max), mean (T_mean) and minimum (T_min).

The soil moisture was maintained at a constant level by daily replenishing lost water with distilled water. A systematic experimental design was established to evaluate the effect of the two amendments used and of the added copper on the plant growth and microbial community response, several experimental conditions were established (Table 1).

Table 1. Experimental conditions and corresponding acronyms.

Conditions	Experimental Acronym
Soil	S
Soil + Cu	S+Cu
Soil + Manure	M
Soil + Manure + Cu	M+Cu
Soil + Digestate	D
Soil + Digestate + Cu	D+Cu

A schematic representation of the experimental design is described in Figure 2.

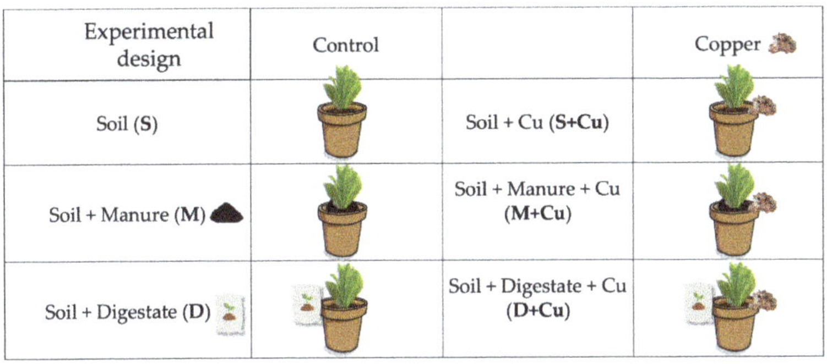

Figure 2. Representation of the experimental design. Lettuce production during five weeks under different soils: soil alone, cattle manure (1% w/w) and digestate (1% w/w), both homogenously added to the soil alone, and the in combination with 30 mg kg^{-1} copper sulphate ($CuSO_4$ $5H_2O$). All treatments were tested in quadruplicates, i.e., twenty-four experimental units were used.

Figure 3 shows images of lettuce plants at harvest for each treatment (a) and the layout of the experiment in the greenhouse investigating lettuce growth in different soil conditions (b).

(a) (b)

Figure 3. (**a**) One representative lettuce plant (side and top view) from each treatment at harvest: S, soil alone; S+Cu, soil and copper; M, manure; M+Cu, manure and copper; D, digestate; D+Cu, digestate and copper and (**b**) The layout of the experiment in the greenhouse for growing lettuce in different soil conditions.

After five weeks, the lettuce plants were harvested and stored for further analyses. Moreover, aliquots of the rhizosphere soil samples were collected for measurements of the microbiological parameters.

2.2. Plant Growth Analyses

At the end of the experiment, the height (cm) of the lettuce plants and the number of leaves (N° leaves) were measured to assess the plant growth. To quantify the plant shoot biomass, the lettuces were cut from the base with scissors and then weighed. The dry weight (DW) of the lettuce samples was obtained by placing leaves in an oven at 60 °C until a constant mass was reached. The leaf dry mass was then related to its leaf area, calculated using image analysis software (ImageJ, National Institutes of Health, Bethesda, MD, USA), to obtain the specific leaf area (SLA, $cm^2\ g^{-1}$ DW). The growth parameters were measured on four biological replicates from each condition (Table 1).

2.3. Determination of Chlorophyll

Photosynthetic pigments of the lettuce plants (0.1 g) were extracted in 96% (*v*/*v*) ethanol for 72 h at room temperature in the dark. After centrifugation at 3000× *g* for 10 min at 4 °C, the absorbance of the supernatant was read spectrophotometrically at 663 and 645 nm (Lambda 35 UV/VIS, Perkin Elmer, Norwalk, CT, USA) [18]. The total chlorophyll content was assessed according to the equations described by Lichtenthaler et al. [19] and the results are expressed in mg of the total chlorophyll per gram of fresh weight plant tissue (mg g^{-1} FW).

2.4. Chlorophyll Fluorescence Analysis

The chlorophyll *a* fluorescence transient (OJIP transients) was measured on the fully expanded leaves of the *Lactuca sativa* plants using a portable fluorometer (FluorPen FP 110-LM/D, Photon System Instruments, Drasov, Czech Republic). The measurements were performed on 60 min dark-adapted leaves with leaf clips, and the fluorescence intensity was measured after the application of a saturating light pulse of 3000 μmol $m^{-2}\ s^{-1}$. These measurements were taken on 10 plants (one leaf per plant) per treatment (S, S+Cu, M, M+Cu, D, and D+Cu), around midday (between 10.00 a.m. and 12.00 p.m.). The obtained data were used in the JIP test [20] to determine various bioenergetic parameters of PSII photochemistry (Table 2).

Table 2. Abbreviations, formulas, and definitions of the JIP-test parameters.

Basic Parameters Calculated from the Extracted Data	
F_0	Fluorescence intensity at 50 μs (O step)
F_{300}	Fluorescence intensity at 300 μs
F_J	Fluorescence intensity at 2 ms (J step)
F_I	Fluorescence intensity at 30 ms (I step)
F_m	Maximal fluorescence intensity (P step)
$F_v = F_m - F_0$	Maximal variable fluorescence
$V_J = (F_J - F_0)/(F_m - F_0)$	Variable fluorescence at J step
$M_0 = 4\ (F_{300} - F_0)/(F_m - F_0)$	Approximated initial slope of the fluorescence transient, expressing the rate of RCs' closure
Specific energy fluxes (per RC: QA-reducing PSII reaction center), in ms^{-1}	
$ABS/RC = M_0 \times (1/V_J) \times [1/(F_v/F_m)]$	Absorption per active reaction center
$TR_0/RC = M_0 \times (1/V_J)$	Trapping per active reaction center
$ET_0/RC = M_0 \times (1/V_J) \times (1 - V_J)$	Electron transport per active reaction center
$DI_0/RC = (ABS/RC) - (TR_0/RC)$	Dissipation per active reaction center
Quantum yields and efficiencies	
$TR_0/ABS = F_v/F_m = \varphi P_0 = (F_m - F_0)/F_m$	Maximum quantum yield of PSII photochemistry
$ET_0/TR_0 = \psi_0 = (F_m - F_J)/(F_m - F_0)$	Probability that a trapped exciton moves an electron into the electron transport chain beyond Q_A
$ET_0/ABS = \varphi E_0 = \varphi P_0 \times \psi E_0$	Quantum yield of electron transport
$F_v/F_0 = TR_0/DI_0 = (F_m - F_0)/F_0 = \varphi P_0/(1 - \varphi P_0)$	Maximum ratio of quantum yields of photochemical and concurrent non-photochemical processes in PSII
$F_0/F_m = DI_0/ABS = \varphi D_0$	Maximum quantum yield for energy dissipation at the antenna level
Performance index	
$PI_{ABS} = [\varphi P_0\ (V_J/M_0)] \times [\varphi P_0/(1 - \varphi P_0)] \times [\psi E_0/(1 - \psi E_0)]$	Performance index (potential) for energy conservation from photons absorbed by PSII to the reduction of intersystem electron acceptors

2.5. Measurement of Lipid Peroxidation Level

Lipid peroxidation was determined by a measurement of the malondialdehyde (MDA) content according to the modified method of Hodges et al. [21]. Frozen samples (0.1 g) were homogenized in a pre-cooled mixer mill (TissueLyser LT, Quiagen, Hielden, German) with 1 mL of 0.1% trichloroacetic acid (TCA) and centrifuged for 10 min at 13,000× *g* at 4 °C. An assay mixture containing 0.4 mL of supernatant, 1 mL of 0.5% (*w*/*v*) thiobarbituric acid (TBA) in 20% (*w*/*v*) TCA, and 1 mM ethylenediamine tetraacetic acid (EDTA) was incubated at 80 °C for 30 min and then quickly cooled in an ice bath. Then samples were centrifuged at 13,500× *g* for 5 min at 4 °C, and the absorbance of the supernatant was measured at 532 nm and 600 nm with a Thermo Multiskan FC Microplate Photometer (Thermo Fisher Scientific; Waltham, MA, USA). Values corresponding to non-specific absorption at 600 nm were subtracted. The MDA concentration was calculated using the extinction coefficient (ε = 155 mM^{-1} cm^{-1}). The lipid peroxidation levels are expressed as nanomoles of MDA per gram of fresh weight plant tissue (mg g^{-1} FW).

2.6. Estimation of Total Phenolic Content

Frozen leaves (0.5 g), homogenized in a pre-cooled mortar and pestle, were extracted in 5 mL of 80% (*v*/*v*) ethanol, placed on a shaker for 24 h in the dark at 25–30 °C, then centrifuged at 4 °C for 25 min at 4000× *g*. The total phenolic content was quantified using Folin–Ciocalteu (FC) reagent, with gallic acid as the standard according to the modified protocol of Bernabé-Antonio et al. [22]. Briefly, 20 µL of extract was mixed with 20 µL of 10% FC reagent and 160 µL of 1M Na_2CO_3. After 20 min of incubation in the dark, the absorbance was measured at 765 nm using a Thermo Multiskan FC Microplate Photometer (Thermo Fisher Scientific; Waltham, MA, USA). The total phenolic content is expressed as mg of gallic acid equivalents (GAE)/g of fresh weight plant tissue (mg GAE g^{-1} FW) using the gallic acid calibration curve (0.1–0.75 mg mL^{-1}, R^2 = 0.97).

2.7. Chemical Analysis

To quantify the chemical element contents in the plants and soils, both oven-dried lettuce leaves and soil samples were weighed and mineralized with a nitric-acid-based microwave digestion method using an EXCEL Microwave digestion system (Preekem Scientific Instruments Co., Ltd., Shanghai, China). Mineralization was performed by treating 250 mg of dried lettuce leaves with 4 mL HNO_3 65% (*w*/*w*), 2 mL H_2O_2 30% (*v*/*v*), and 3 mL of distilled water. Regarding the metal concentration in the soil, 0.2 g of dried soil samples was digested with 5 mL of HNO_3 65% (*w*/*w*) and 1 mL of HCl 37% (*v*/*v*). Samples were then filtered and analyzed. The determination of elements was performed using inductively coupled plasma optical emission spectrometry at 327.395 nm (ICP-OES, Agilent 5800, USA-LOD = 0.02 mg L^{-1}).

2.8. Bioaccumulation Factor of Copper in Lettuce Plants

The bioaccumulation factor (BAF) was calculated to evaluate the ability of the lettuce plants to extract and accumulate copper from the soil to shoot [23]:

$$\text{BAF} = \text{Cu content in plant shoot (mg kg}^{-1}\text{ DW)/Cu content in soil (mg kg}^{-1}\text{ DW)} \quad (1)$$

2.9. Tolerance Index of Lettuce Plants toward Copper

The tolerance index (Ti) was measured to assess the growth of the lettuce plants in Cu-contaminated soil (Cu) compared to the control soil (S) [24]:

$$\text{Ti \%} = (FW_{Cu}/FW_S) \times 100 \quad (2)$$

where FW_{Cu} is the fresh weight (g) of shoot lettuce grown in Cu-contaminated soil and FW_S is the fresh weight (g) of shoot lettuce grown in non-contaminated soil.

2.10. Microbiological Analyses of Soil Community

Microbiological analyses of the soil community were performed at the end of the experiment, after 36 days. The total microbial abundance was measured in the soil samples (1 g), previously fixed in 2% formaldehyde (v/v), by the direct epifluorescence count method with a Leica DM LB 30 epifluorescence microscope and using DAPI (4′,6-Diamidino-2-phenylindole dihydrochloride) dye as a fluorescent DNA intercalant [25,26]. The total microbial abundance is expressed as N. cells g^{-1} dry soil.

Microbial activity was detected by measuring the dehydrogenase activity (DHA) in the soil samples (6 g) using tetrazolium salt, 2,3,5-triphenyl-tetrazolium chloride (TTC) as an artificial terminal hydrogen acceptor in the electron transport chain. TTC was reduced to red-colored triphenylformazan (TPF). TPF was extracted using an organic solvent (ethanol), and the color intensity of the extract was quantified after 24 h at 37 °C in the dark [27] with a Thermo Multiskan FC Microplate Photometer (Thermo Fisher Scientific; Waltham, MA, USA). The intensity of the color was directly proportional to the concentration of the produced TPF. Soil dehydrogenase activity is expressed as μg TPF g^{-1} dry soil.

2.11. Statistical Analysis

All collected data were statistically analyzed by one-way analysis of variance (ANOVA) using R software (version 4.1.2 https://www.r-project.org; accessed on 4 December 2023). Differences among the different treatment means were evaluated by post hoc Tukey's Honest Significant Difference test (Tukey HSD) with a significance level of $p < 0.05$. In the figures, the results are presented as box plots with the median value displayed as a line in the middle of the box. Maximum and minimum observations are shown as the endpoint of the upper and lower whiskers, respectively, and outliers are plotted as dots when present. The box plots were made using R ggplot2 package (https://cran.r-project.org/web/packages/ggplot2/citation.html; accessed on 4 December 2023), In the tables, the results are expressed as mean values ± standard deviations (n = 4). The relationships among the plant growth, biochemical parameters, and microbiological data in the soil were analyzed using Principal Component Analysis (PCA).

3. Results and Discussion

3.1. Plant Growth Analyses

The effects of different combinations of cattle organic soil amendments and copper on the lettuce plants growth were analyzed by evaluating the shoot biomass production (Figure 4). In the presence of manure (M) and digestate (D), the plant biomass increased significantly (by 30% and 140%, respectively) compared to the control plants (S). It is worth noting that the lettuces grown on D soil showed the highest shoot biomass. A comparison across separate studies may be quite difficult due to the influence of several variables such as different compositions of biochar, manure or digestate, tested plants and soil, and pot experiment conditions. Nevertheless, our findings were in accordance with Kwoczynski et al. [6], who reported a two-fold increase in the yield of lettuce plants (*Lactuca sativa* L.) grown in the presence of 5% (w/w) digestate compared to controls [6]. Similarly, the addition of digestate to soil caused an enhancement in biomass yield also in eggplant (*Solanum melongena* L.) [28], ryegrass (*Lolium perenna* L.) [29], tomato (*Solanum lycopersicum* L.) [30], and forage grass (*Lolium rigidum* L.) [31], resulting in an important source of nutrients. The increase in lettuce shoot biomass might be associated with a slow release of nutrients from the digestate that occurred during the development of the crop, as suggested by Mortola et al. [5]. The addition of a low Cu concentration to the soil (S+Cu) and manure (M+Cu) resulted in a significant reduction in shoot biomass compared to the soil (S) and manure (M) (32.1% and 30.9%, respectively), highlighting a harmful impact of copper sulphate (30 mg kg^{-1}) on biomass production. Instead, no significant difference was shown in the shoot biomass of the lettuce plants grown in the presence of digestate with or without copper. These results are in line with those reported by Gharbi et al. [32] and Shams et al. [33], who observed a decline in the biomass production of lettuce plants exposed to increasing Cu concentrations (100–1000 mg kg^{-1}) for 15 days and 50 and 100 mg kg^{-1} Cu for 30 days, respectively.

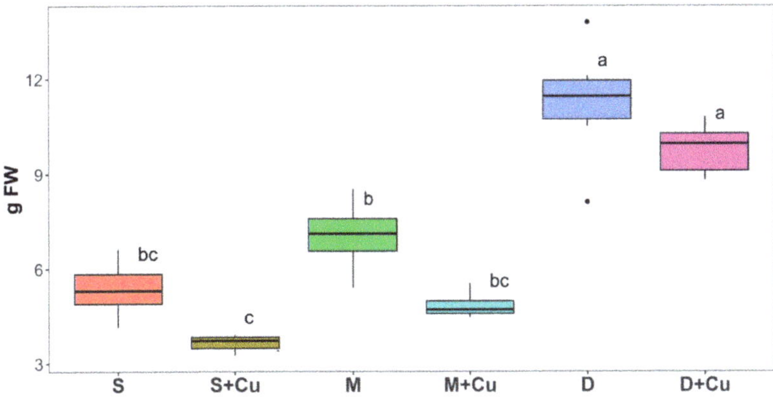

Figure 4. Shoot biomass (g FW) of lettuce plants grown for five weeks in soil (S), manure (M), and digestate (D) and with the addition of copper (Cu). Box plots labelled with different letters indicate significant differences between different treatments (ANOVA followed by Tukey HSD, p value ≤ 0.05). In the box-plots, the upper and lower whisker endpoints represent the maximum and minimum observations, respectively, outliers are shown as dots. $n = 4$.

No significant difference in plant height was observed between the soil (S), manure (M), and digestate (D) treatments. A significant downward trend in plant height was detected in S+Cu and M+Cu, at 19.12% and 11.36%, respectively, compared to the S and M conditions. Instead, the addition of copper to digestate significantly increased the height of the lettuce plants (3.8%) (Table 2). The number of leaves significantly increased in the presence of M and D, at 40.9% and 46.5%, respectively, with respect to the lettuce plants grown in the S. Copper addition to S, M, and D did not cause changes in the numbers of leaves compared to the respective uncontaminated soil improvers (Table 3). It is to be noted that the numbers of lettuce leaves in S+Cu were lower than those detected for the plants grown in M+Cu and D+Cu. Considering the specific leaf area (SLA) parameter, no significant difference was observed between the S, M, and D conditions. Accordingly, the addition of copper did not induce any effect on the SLA values (Table 3). The analyzed growth parameters (height, number of leaves, and SLA) were consistent with the previously measured biomass results. Previous studies have indicated that the use of soil organic fertilizer contributes to the better development of crop plants such as lettuce [14], eggplant, Shanghai cabbage [34], and tomato [35]. This might be ascribed to the high nutrient and organic matter contents of amendments, which optimize both the deficit and excess nutrients in the soil [4]. The addition of a low copper amount adversely affected lettuce growth both in the soil and in manure, interfering with plant development and associated metabolic processes [10].

Table 3. Shoot height (cm), leaf density (N° leaves), and specific leaf area (SLA, cm^2 g^{-1} DW) for lettuces plants collected from different soil conditions: S, M, and D; and in presence of copper (Cu): S+Cu, M+Cu, and D+Cu.

	S	S+Cu	M	M+Cu	D	D+Cu
Height	14.6 ± 1.2 ^{ab}	11.8 ± 0.8 ^{c}	14.3 ± 1.0 ^{ab}	12.9 ± 1.9 ^{bc}	14.5 ± 1.2 ^{ab}	15.1 ± 1.0 ^{a}
N° leaves	11.9 ± 1.1 ^{b}	11.1 ± 0.4 ^{b}	16.5 ± 0.5 ^{a}	15.9 ± 1.2 ^{a}	17.4 ± 0.5 ^{a}	16.3 ± 1.1 ^{a}
SLA	2.9 ± 0.2 ^{a}	3.0 ± 0.5 ^{a}	2.6 ± 0.9 ^{a}	2.9 ± 0.4 ^{a}	3.0 ± 0.4 ^{a}	2.9 ± 2.0 ^{a}

Means $\pm$ SD with different letters in the vertical columns indicate significant differences between different treatments (ANOVA followed by Tukey HSD, p value ≤ 0.05) ($n = 4$).

3.2. Chlorophyll Content

A downward trend in the chlorophyll content was observed under the M (9%) treatment compared with S (Figure 5). The addition of copper significantly reduced the chlorophyll content both in S+Cu and M+Cu (18,48% and 12.48%, respectively) compared to their respective uncontaminated soil improvers. A reduction in the photosynthetic pigment content suggested a significant inhibition of photosynthesis, thus affecting the plant growth, as reported by Slamet et al. [36] Indeed, the pattern observed in the plant biomass production (Figure 5) was consistent with that concerning the chlorophyll content (Figure 5). The addition of copper to the soil and manure induced chlorophyll and lettuce biomass reductions, even if the Cu concentration was lower than that reported in the literature (from 200 mg kg^{-1}), which causes toxicity [33]. In agreement with the study by Li et al. [37], low copper concentrations (25–50 mg kg^{-1}) might induce a hormesis effect and a consequent reduction in biomass and pigment contents. The chlorophyll content was significantly increased in lettuce plants grown with D (32%) with respect to S (Figure 5). No significant difference was highlighted in the chlorophyll content measured in the lettuce plants grown with digestate in the absence and presence of copper. The increase in chlorophyll content in the lettuce plants grown with digestate and copper was 30% with respect to S (Figure 5). Chlorophyll is an essential pigment for plant growth and a higher chlorophyll content is usually linked to an increase in plant biomass accumulation. Digestate contains available nitrogen and phosphorus, therefore, its use can stimulate photosynthesis, thus enhancing the chlorophyll content and biomass in a range of vegetable crops, including lettuce (*Lactuca sativa* L.) [14], tomato (*Solanum lycopersicum* L.) [38], Spring Wheat (*Triticum aestivum* L.) [39], and aubergine (*Solanum melongena* L.) [28].

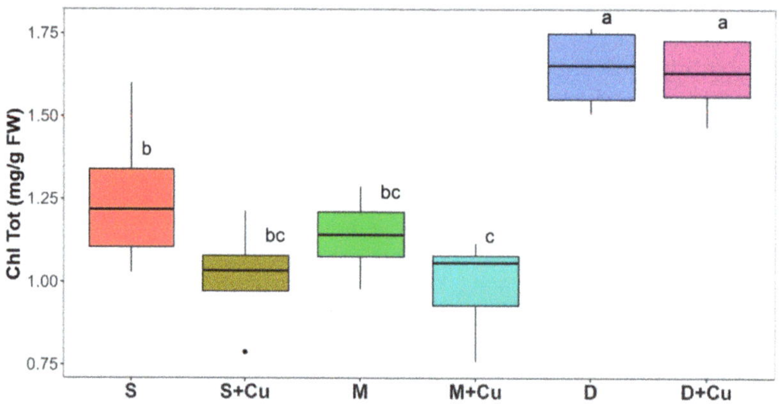

Figure 5. Total chlorophyll content (mg g^{-1} FW) of lettuce plants grown for five weeks in soil (S), manure (M), and digestate (D) and with the addition of copper (Cu). Box plots labelled with different letters indicate significant differences between different treatments (ANOVA followed by Tukey HSD, p value ≤ 0.05). In the box-plots, the upper and lower whisker endpoints represent the maximum and minimum observations, respectively, outliers are shown as dots. $n = 4$.

3.3. Chlorophyll Fluorescence Analysis

In order to study the effect of the soil amendments on the photosynthetic performance of the lettuce plants exposed to a low Cu concentration, the bioenergetic parameters obtained from the JIP test were measured. In particular, some structural and functional parameters are presented in the form of a radar plot (Figure 6), where all values have been normalized to those obtained from the control plants.

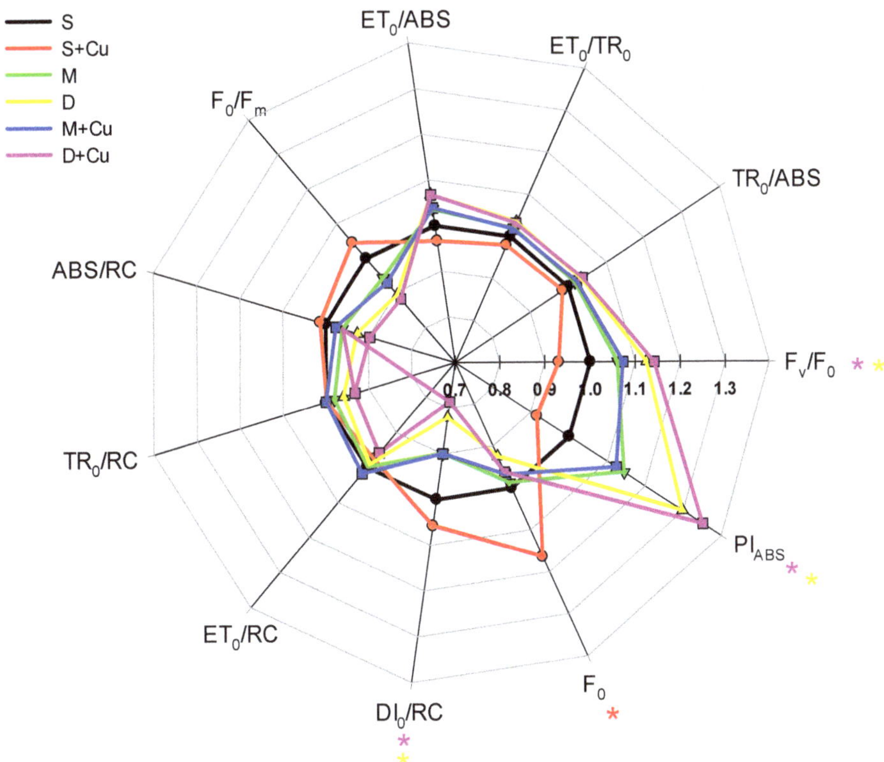

Figure 6. Changes in the shape of the radar plot JIP-test parameters induced by different amendment treatments with or without the addition of a low copper concentrations applied to *Lactuca sativa* plants and measured at the end of the treatment. The data are the average of ten replicates and report the values with respect to plants grown in the control condition (S = control = 1). Red, pink, and yellow asterisks indicate significant differences (LSD test, $p < 0.05$) between control plants (S) and those exposed to S+Cu, D, or D+Cu, respectively.

A chlorophyll a fluorescence-transient analysis is a useful method for examining the physiological aspects of structure and activity, particularly in the PSII [20]. It has been extensively employed to assess damage to the plant photosynthetic system caused by various types of stress [40]. The exposure of lettuce plants to a low copper concentration, as in S+Cu, caused a slight change in most of the parameters analyzed, except for F_0, which exhibited a significant increase compared to the control (S) plants (Figure 6). As reported in the literature, the absence of an F_0 variation indicates a good ability of the treated plants to maintain the efficiency of energy transfer between the pigments of the antenna and the PSII reaction center without structural damage at the photosystem level [41]. However, an increase in F_0 can be interpreted as indicating irreversible damage to PSII caused by uncontrolled dissipation of heat that produces an excess of excitation energy [42,43], while a decrease in F_0 is, on the contrary, an indication of a high-energy dissipation in the minor antenna [44]. The data relating to the specific energy fluxes, such as the absorption flux per reaction center (ABS/RC), trapped energy flux per reaction center (TR_0/RC), electron transport flux per reaction center (ET_0/RC), and dissipated energy flux per reaction center (DI_0/RC), measured to determine the photosynthetic performance of the active PSII reaction centers of the lettuce plants subjected to Cu treatment (S+Cu), did not show significant variations with respect to the control plants (Figure 6). Nevertheless, although copper treatment was found to not significantly impact the other analyzed parameters in the S+Cu plants, such as the maximum quantum yield of primary photochemical reactions (TR_0/ABS), the maximum quantum yield for electron transport (ET_0/ABS), the probability that an electron moves further than Q_A (ET_0/TR_0), and the light energy used for thermal energy dissipation (F_0/F_m), both the (F_v/F_0) and the performance index PI_{ABS} showed a slight decrease compared to the control, indicating a reduction in the efficiency of water splitting by the PSII and the overall vital state of the plant [45].

The exposure of lettuce plants to manure amendments in the soil, without or with the addition of a low copper concentration (M or M+Cu, respectively), resulted in a partial improvement in the plant physiological status. Compared to the plants grown in the S+Cu treatment, some of the analyzed parameters in the plants grown under the M and M+Cu treatments showed a positive impact on the plant viability but did not significantly differ from the control plants. These results are consistent with the biomass and chlorophyll content data, which indicate similar grown patterns across the treatments (S, S+Cu, M, and M+Cu).

On the contrary, the exposure of the lettuce plants to digestate amendments in soil, without or with the addition of a low copper concentration (D or D+Cu, respectively), significantly enhanced the plant physiological performance.

Several chlorophyll fluorescence parameters measured in the plants grown under the D or D+Cu treatments were found to be significantly different from those of the control plants (S), indicating a beneficial effect on the plant viability due to the digestate amendment.

Indeed, in our study, the functionality of both photosystems I and II was noticeably higher in the D- and D+Cu-treated lettuce plants, as indicated by the increase in the performance index PI_{ABS} and in the F_v/F_0 values compared to the control [14]. Additionally, the decrease in the dissipated energy flux per reaction center (DI_0/RC) indicated a greater ability to use the energy by the active RCs [46]. The data on the biomass and chlorophyll content confirm these findings, showing an increase in biomass production and pigment content in the plants treated with D or D+Cu compared to the controls.

The results obtained in this experiment are consistent with previous studies that demonstrate the positive impact of digestate addition on crop growth [47–49]. Notably, significant increases in biomass have been reported for wheat (*Triticum aestivum* L.) [50], giant Napier (*Pennisetum purpureum* Schumach) [51], tomato (*Solanum lycopersicum* L.), and cucumber (*Cucumis sativus* L.) [52].

3.4. Measurement of Lipid Peroxidation Levels

The malondialdehyde (MDA) content increased significantly in the manure (M) and digestate (D) conditions (by 32% and 86%, respectively) compared to the control condition (S), (Table 3). Malondialdehyde is a natural by-product of cell membrane lipid peroxidation, usually analyzed to assess the damage caused by oxidative stress in plant tissues [14]. The addition of low concentrations of copper significantly reduced the MDA content in the lettuces grown in S+Cu (22%) compared to soil (S). In the M+Cu condition, the measured MDA values were significantly reduced (66%) when compared with M, and no significant difference was detected compared to the control (S), highlighting a partial improvement of the plant physiological performance, strictly consistent with previous results concerning the chlorophyll fluorescence analysis. The MDA content in the lettuces planted in D+Cu did not differ significantly from the digestate (D) conditions, maintaining high MDA levels. If MDA levels persist to be high, irreversible modifications to proteins and nucleic acids occur. However, if the scavenging system composed of enzymatic and non-enzymatic systems synergistically works to maintain intra- and extracellular redox homoeostasis, MDA increases, and its maintenance may represent an acclimation process rather than damage, as suggested by Morales et al. [53] and Liang et al. [54]. Consequently, MDA might act as a protection factor exerting a positive role by activating the regulatory genes involved in plant defense and development according to Tagnon et al. [55]. Additional parameters, such as growth measurements and chlorophyll fluorescence analyses, in plants grown under the D and D+Cu treatments showed a beneficial effect on plant vitality due to the presence of digestate when compared to the S and M conditions with and without copper.

3.5. Estimation of Total Phenolic Content

Lettuces of different varieties have different total phenol contents due to genetic characteristics and different agronomic practices, such as fertilizer application, which could influence the nutritional status of the crop, modulating the antioxidant and phenol contents [56]. Phenols are secondary metabolites used by plants, along with antioxidants, to defend against oxidative or abiotic stresses [28]. Lettuces grown with digestate (D) and digestate plus copper (D+Cu) showed the highest content of total phenolic content, (20% and 23% respectively) compared to the control condition (S). The phenolic content in the lettuce plants did not change significantly between the different soil conditions, with or without copper addition (Table 4). These findings agreed with the results obtained for cucumber [57] and lettuce plants [58], where organic fertilizers enhanced the total phenol content, since they are rich in carbon and phenols, and stimulated plant antioxidant activity. Phenols can be found in organic fertilizer from the biological degradation of several xenobiotic substances, like pesticides, or naturally occurring aromatic amino acids and polymers in plant materials such as lignin, tannins, and humic acids [28].

Table 4. Measurement of lipid peroxidation levels (MDA, nmol MDA g^{-1} FW) and estimation of total phenolic content (TPC, mg GAE g^{-1} FW) for lettuces plants collected from different soil conditions: S, M, and D; and in presence of copper (Cu): S+Cu, M+Cu, and D+Cu.

	S	S+Cu	M	M+Cu	D	D+Cu
MDA	5.0 ± 0.6^{c}	3.9 ± 0.4^{d}	6.6 ± 0.7^{b}	4.3 ± 0.5^{cd}	9.3 ± 0.5^{a}	9.5 ± 1.3^{a}
TPC	0.30 ± 0.05^{ab}	0.32 ± 0.07^{a}	0.31 ± 0.01^{a}	0.35 ± 0.03^{a}	0.36 ± 0.07^{a}	0.37 ± 0.05^{a}

Means ± SD with different letters in the vertical columns indicate significant differences between different treatments (ANOVA followed by Tukey HSD, p value ≤ 0.05). $n = 4$.

3.6. Copper Determination, Bioaccumulation Factor, and Tolerance Index in Lettuce

Copper is an essential element for plant nutrition, needed as a cofactor for many enzymatic activities catalyzing redox reactions and for photosynthetic functions. Excess copper induces high levels of reactive oxygen species (ROS) and negatively affects the photosystem during photosynthesis [10]. About Cu concentration in soil any differences were detected in S, M and D (Table S2). Although copper addition increased its concentration in different soils, no significant difference was observed between them (Table S2). The lettuce plants grown under the control condition (S) showed the lowest Cu accumulation (1.1 ± 0.2 mg kg^{-1}) (Table 5). Compared to the S condition, in lettuce plants grown with M and D, a marked increase in Cu concentration was observed, following the order of D > M > S. The addition of a low Cu concentration to different soils (S+Cu, M+Cu, and D+Cu) induced a significant enhancement in copper accumulation in the lettuce plants, following the order of M+Cu > D+Cu > S+Cu > S.

Table 5. Copper content in leaves (Cu, mg kg^{-1} DW), bioaccumulation factor (BAF) of Cu (BAF_{Cu}), and tolerance index (Ti) for lettuces plants collected from different soil conditions: S, M, and D; and in presence of copper (Cu): S+Cu, M+Cu, and D+Cu.

	S	S+Cu	M	M+Cu	D	D+Cu
Cu	1.1 ± 0.1^{c}	3.3 ± 0.3^{b}	2.9 ± 0.4^{b}	5.7 ± 0.4^{a}	4.2 ± 0.2^{b}	3.9 ± 0.2^{b}
BAF_{Cu}	0.04 ± 0.01^{c}	0.14 ± 0.06^{ab}	0.23 ± 0.09^{a}	0.19 ± 0.07^{ab}	0.15 ± 0.03^{ab}	0.10 ± 0.03^{ab}
Ti%	100 ± 0.00^{b}	53 ± 7^{d}	105 ± 6^{b}	77 ± 13^{c}	134 ± 10^{a}	121 ± 2^{a}

Means ± SD with different letters in the vertical columns indicate significant differences between different treatments (ANOVA followed by Tukey HSD, p value ≤ 0.05). $n = 4$.

Based on the Cu contents in the plant shoot and soil, the bioaccumulation factor (BAF) and tolerance index (Ti) were calculated to evaluate the metal uptake and plant tolerance to Cu stress. The highest BAF was detected in the lettuce plants grown with M soil (0.23 ± 0.09 mg kg^{-1}) compared to the different uncontaminated soils (Table 5). The addition of a low Cu concentration caused a significant enhancement in the BAF value in the lettuces grown with S+Cu compared to the S condition, not statistically different from those occurring in the M+Cu and D+Cu soils. It is well known that the BAF value depends on the tested metal, plant species, soil pollution level, and soil acidity, as reported by Sipter et al. [58,59]. Nevertheless, in this study, the BAF value in the plant leaves was <1, revealing that the lettuces were poor accumulators [60,61] (Table 5). The plant capability to limit Cu's damaging effects was very different among soil conditions, as evidenced by the tolerance index (Ti) (Table 5). The lettuce plants grown in D+Cu soil showed the highest Ti values with respect to the S+Cu and M+Cu conditions, even though the respective BAF values were similar. These results are consistent with data concerning plant growth and physiological parameters, suggesting that the presence of digestate in the soil reduced Cu's phytotoxic effect, improving the plant physiological status. It is worth noting that the lowest Ti was observed in the lettuces planted in S+Cu, which corresponded to a 3.5-fold increase in the BAF value compared to the S conditions and a decrease in the growth parameters and chlorophyll content. Although the Cu concentration at which toxic effects become evident is plant-species- and soil-type-dependent [62], our findings agree with previous studies where plants with Ti values of <100 experienced stress due to metal contamination along with a net reduction in plant biomass. Conversely, the plant species with a T_i of >100 developed tolerance along with a net biomass increase and improved physiological performance [63,64].

3.7. Macro- and Micro-Nutrient Content in Lettuce Plants

It is well established that plant growth and nutritional quality are related to mineral accumulation. Macronutrients such as Ca, K, Mg, P, and S carry out important functions in several metabolic pathways, including photosynthesis, growth, and stress tolerance, and are required in large quantities by plants [65]. Compared to macronutrients, micronutrients (Cu, Fe, Mn, Na, Zn, Al, Cd, Cr, Ni, and Si) are required in relatively lesser amounts and, likewise, they play key roles in plant growth and metabolism [66]. The levels of chemical elements recorded in the lettuce plants used in this study were all below the WHO safe limit of the selected heavy metals in vegetables for human consumption [67]. Macro- and micronutrient uptake in the lettuce leaves was reported as a radar plot (Figure 7), where all values were normalized to those obtained from the control plants (S) (Table S4). In general, the uptake of macro- and micronutrients varied significantly in the lettuce plants treated with different soil amendments in the absence or presence of a low Cu concentration, resulting to be higher in plants grown in S+Cu. Compared to the control plants (S), the presence of the amendments mainly affected the uptake of micronutrients, such as Cr, Ni, Cu, and Cd. In particular, under the M treatment, the leaves of lettuce plants showed reductions to 36.7% and 41.7% in Ni and Cd concentrations, respectively, whereas an enhancement was observed in Cr and Cu uptake (up to 45.7% and 91%, respectively) with respect to the control (S). Instead, in the lettuce plants exposed to D, marked reductions were detected in Ni and Cd uptake (up to 50% and 66%, respectively) whereas Cr and Cu concentrations increased to 25% and more than twice, respectively, compared to the control (S). It is worth noting that, under the D and D+Cu treatments, the plants showed significant decreases also in Al, Zn, Na, Mn, and Fe uptake (13–16%, 14–20%, 10–12%, 14–17%, and 17–19%, respectively). As described in the literature, the uptake of inorganic nutrients and metals is dependent on both the metal and soil, and strongly influenced by the amount of water transpired by the plant. As a consequence of the rise in temperature and the subsequent increase in transpiration, there is the potential for an increase in the transfer of copper from the soil to the plant [16]. The reduction in micronutrients in the lettuce leaves could likely be ascribed to their lower solubility due to the higher digestate pH compared to the soil one (Table S1), as reported by Olszyk et al. [68], who observed a decrease in Fe, Mn, and Zn concentrations, mainly, in lettuce plants grown in the presence of several biochars as amendments. Similarly, a reduction in Fe uptake was detected in baby leaf lettuce under different digestate treatments [69,70]. The addition of a low Cu concentration (M+Cu and D+Cu) caused in the plant leaves a two-fold increase in Cu uptake with respect to the control (S). Moreover, as reported previously, our findings highlighted that the Cu concentration was higher in the plants grown in M+Cu compared to those exposed to S+Cu and D+Cu.

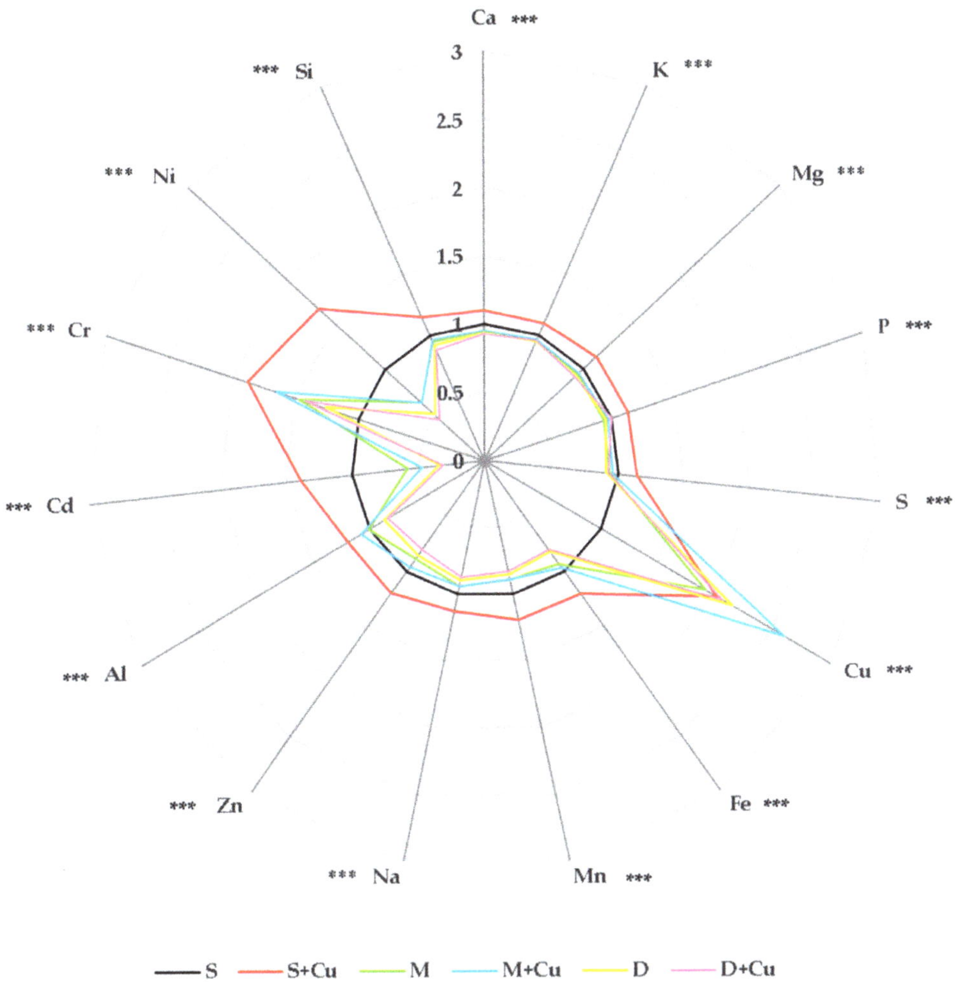

Figure 7. Changes in the shape of the radar plot relative to plant uptake of macro- and micronutrients induced by different amendment treatments with or without the addition of a low copper concentrations applied to lettuce plants and measured at the end of the treatment. The data are the average of four replicates and report the values with respect to plants grown in the control condition (S = control = 1). Asterisks indicate significant differences (Tukey HSD, $p < 0.05$) between control plants (S) and those exposed to M, M+Cu, D, and D+Cu.

3.8. Microbiological Analyses of Soil Community

At the end of the experiment, after five weeks, the total microbial abundance was not significantly different between the tested soils (S, M, and D) (Figure 8a). The addition of a low Cu concentration significantly reduced the microbial abundance in the S+Cu soil (36%) whereas a significant enhancement was observed in the M+Cu soil (35%) compared to S and M, respectively. The higher microbial abundance in M+Cu with respect to S+Cu could be ascribed to the presence of copper-resistant bacteria in the manure [71]. Moreover, organic soil amendments could immobilize metals, reducing their bioavailability, due to the presence of humic acids able to chelate a wide variety of metals, including copper [72]. No significant difference in microbial abundance was highlighted between the D and D+Cu soils.

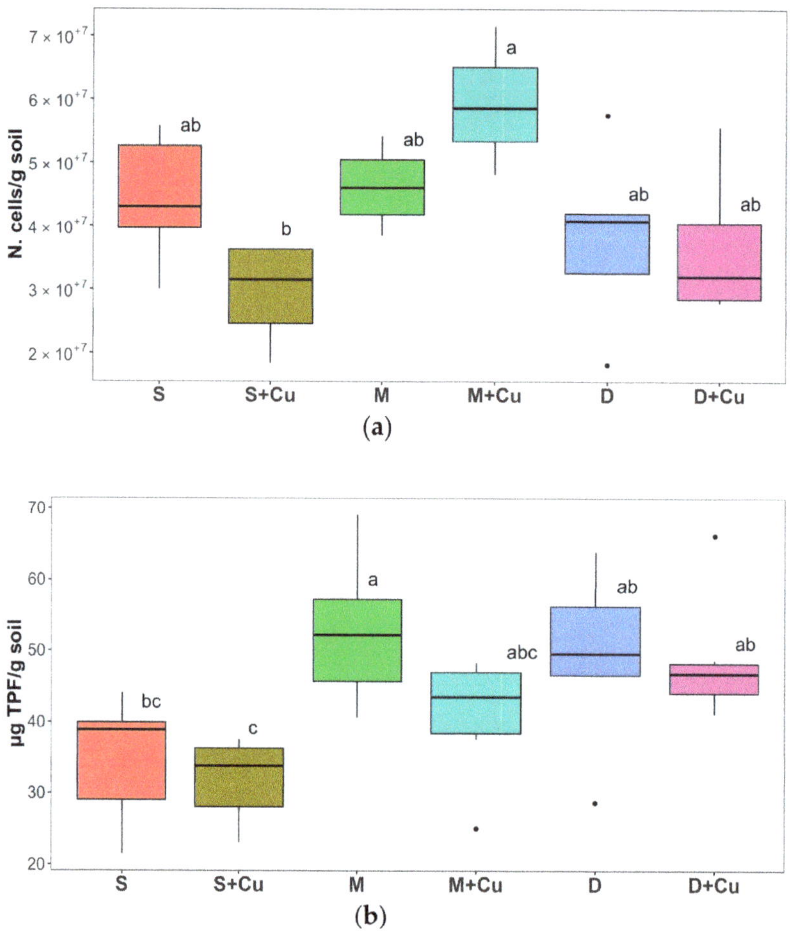

Figure 8. (**a**) Soil total microbial abundance (N. cells g^{-1} dry soil) and (**b**) soil microbial activity (ug TPF g^{-1} dry soil) measured in soil (S), manure (M), and digestate (D) and with the addition of copper (Cu) after five weeks of experimentation. Box plots labelled with different letters indicate significant differences between different treatments (ANOVA followed by Tukey HSD, p value ≤ 0.05). In the box-plots, the upper and lower whisker endpoints represent the maximum and minimum observations, respectively, outliers are shown as dots. $n = 4$.

Both cattle organic soil amendments, manure (M) and digestate (D), showed higher soil microbial activity, 51.24% and 41.31%, respectively, when compared to the non-amended soil (S). Following Cu addition (S+Cu, M+Cu, and D+Cu), a downward trend in dehydrogenase activity was observed compared to the respective soils without amendments (S, M, and D, respectively) (Figure 8b). Dehydrogenase activity (DHA) is a biomarker related to soil fertility and measures the metabolic state of soil microbial communities. In addition, this enzyme reflects the total range of oxidative activity of soil microorganisms [73]. Soil microflora contributes to the improvement of plant physiological performance status. Any application of a toxicant that might affect the growth of soil microorganisms can induce alterations in the general activity of enzymes, such as dehydrogenase [74]. As reported in previous works, DHA activity is increased by the addition of organic amendments in the soil [73], such as the use of manure to fertilize winter rye, potato, oats, barley, and red clover in crop rotations. Heavy metals, including copper, might negatively affect soil bacterial communities by reducing their populations, changing their structure and diversity, decreasing soil enzyme activity, and interfering in plant–soil–metal associations [70].

3.9. Principal Component Analysis

The first two components of the PCA analysis explained 68.3% of the overall variance in the lettuce dataset (Dim 1: 43.1%; Dim 2: 25.2%) (Figure 9). The contribution (contrib) values of the plant performance parameters showed that shoot biomass, MDA, total chlorophyll, number of leaves, DHA, and DAPI were well represented in Dim1 and Dim2, in contrast with SLA, TPC, and the copper content in leaves (Figure 9a). A correlation analysis of the plant performance parameters (Figure S1) demonstrated that shoot biomass, MDA, total chlorophyll, number of leaves, and DHA were significantly positively correlated with Dim1, while SLA, TPC, DAPI, and leaves' copper content were significantly negatively correlated with Dim1. A correlation analysis of Dim2 revealed that TPC, DAPI, DHA, and leaves' number correlated positively, whereas SLA, copper content in leaves, shoot biomass, MDA, and Chl Tot were negatively correlated. The outcomes of the PCA clearly discriminated the samples according to soil composition, showing six clusters, which reflected the different soil conditions plus the control (S) (Figure 9b). The profile of the lettuce plants grown in D and D+Cu clustered in a Dim1-positive direction, while those grown in S, S+Cu, and M+Cu clustered in a Dim1-negative direction. Plants grown in D and D+Cu were separated from the control (S) and manure (M) plants grown with or without copper towards number of leaves, shoot biomass, Chl tot, MDA, and DHA. Similarly, separation of the control (S) and manure (M) plants grown with or without copper was observed towards TPC, copper content in leaves, SLA, and DAPI. Treatment with D and D+Cu resulted in a better physiological performance in the lettuce plants with respect to the manure treatments and control conditions. The second axis (Dim2) was most closely related to DAPI and TPC, and the samples segregation on Dim2 showed that manure had the highest microbial abundance.

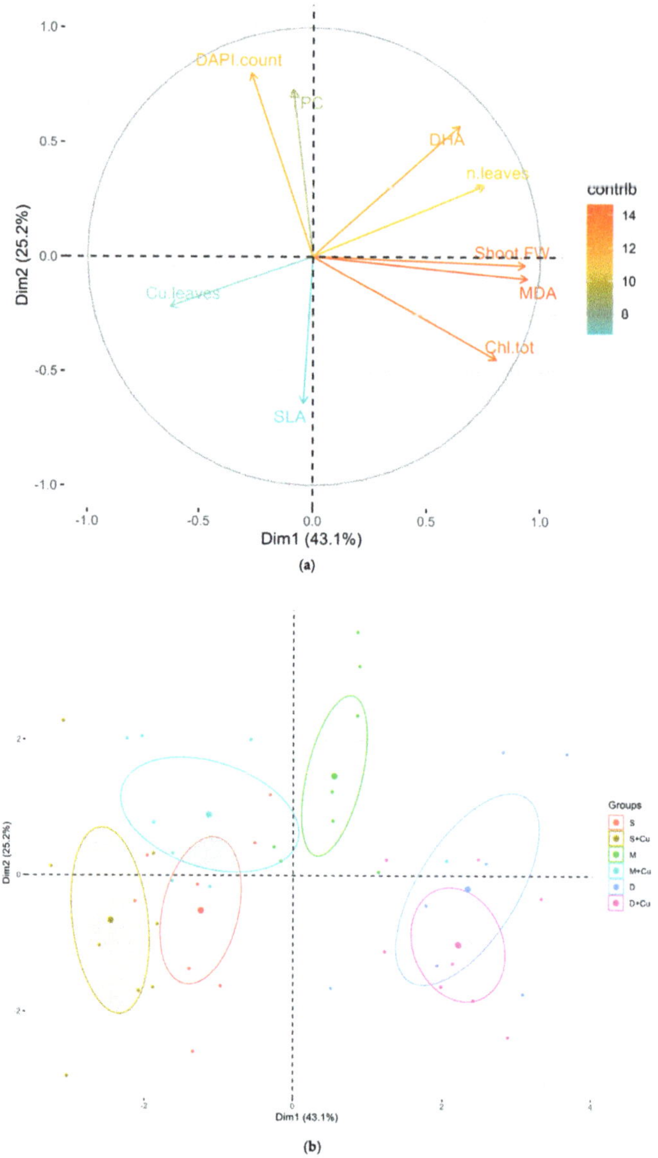

Figure 9. (**a**) Variable correlation plot, constructed using the following variables: shoot fresh biomass (Shoot FW), total chlorophyll content (Chl Tot), specific leaf area (SLA), malondialdehyde content (MDA), copper content in leaves (Cu leaves), leaves number (n.leaves), total phenolic content (TPC), soil total microbial abundance (DAPI count), and soil microbial dehydrogenase activity (DHA). The length of each vector represents the contribution of the corresponding variable on the principal component (Dim1 and Dim2), while the direction of each vector indicates the correlation between the variable and the principal components. Vector colors represent the contribution of the corresponding variable to the overall variability. (**b**) Principal component analysis (PCA) showing samples' segregation by soil composition (S, S+Cu, M, M+Cu, D, and D+Cu). Data points are grouped into colored ellipses representing a 95% confidence interval. Enlarged points within ellipses represent the centroids of the sample distributions grouped by soil composition. The legend provides a color code for each group.

4. Conclusions

This study investigated the effects of growing lettuce plants with different combinations of cattle organic soil amendments, cattle manure (M) or digestate (D), and low concentrations of copper, like those found in agricultural soils used for the cultivation of horticultural species treated with, i.e., copper-based fungicides. The results showed that both manure and digestate affected the lettuce plants very differently. While manure showed negative effects overall, the application of digestate resulted in an increased yield and improved the photosynthetic performance. Digestate application might have buffered the decrease in soil pH caused by fertilization, potentially mitigating soil acidification compared to manure amendment. In the M and M+Cu conditions, a decreased lettuce growth, a higher Cu bioaccumulation, and a reduced tolerance index were associated with increasing the Cu concentration in the soil, suggesting an increased level of Cu phytotoxicity in the plants and a detrimental impact compared to the plants grown in digestate and in the digestate-copper amended soil.

Our greenhouse pot study aligns with existing research on digestate's potential, but field trials are necessary for real-world applicability.

In low- and middle-income countries, digestate offers a more advantageous alternative to manure for replacing chemical fertilizers. As a renewable and inexpensive source of nutrients and energy derived from organic materials, it promotes a circular economy. Large-scale validation is needed to confirm the benefits of anaerobic digestion for a circular economy, as shown by research on energy and fertilizer production from various organic materials.

Supplementary Materials: The following supporting information can be downloaded at: https://www.mdpi.com/article/10.3390/environments11070134/s1; Table S1: Main physico-chemical characteristics of soil, manure and digestate; Table S2: Elements content (mg kg^{-1}) in different soils; Table S3: Elements content in leaves of lettuce plants grown in different soil conditions; Figure S1: Correlation between the variable and the principal components in PCA analyses.

Author Contributions: Conceptualization, P.G., A.B.C. and M.A.I.; methodology, P.G., A.B.C. and M.A.I.; validation, C.D.C. and M.A.I.; formal analysis, C.D.C., A.N., B.C., F.P., D.G. and V.I.; investigation C.D.C., A.N. and F.P.; resources, P.G., A.B.C., F.P. and M.A.I.; writing—original draft preparation, C.D.C., F.P., V.I. and M.A.I.; writing—review and editing, C.D.C., A.N., B.C., F.P., V.I., P.G., A.B.C. and M.A.I.; visualization, C.D.C., F.P., V.I. and M.A.I.; supervision, M.A.I.; funding acquisition, A.B.C. and M.A.I. All authors have read and agreed to the published version of the manuscript.

Funding: The author(s) declare financial support was received for the research, authorship, and/or publication of this article. This work was carried out within the following research initiatives: (i) the Italian Ministry of University and Research, PON "Ricerca e Innovazione" 2014-2020 DM 1061/2021 Azione IV.5 "Dottorati su tematiche Green" (PhD scholarship awarded by Sapienza University of Rome to CDC); (ii) "Gruppi di Ricerca 2020" SMART-BREED Project A0375E0166 (POR FESR LAZIO 2014-2020); (iii) Agritech National Research Center, European Union Next-Generation EU (PIANO NAZIONALE DI RIPRESA E RESILIENZA (PNRR)—MISSIONE 4 COMPONENTE 2, INVESTIMENTO 1.4—D.D. 1032 17/06/2022, CN00000022)—SPOKE 1, Task 1.2.3.

Data Availability Statement: All the data supporting the reported results can be found in this article.

Acknowledgments: We thank: the Orto di Fabiana company (Rome, Italy) for providing *L. sativa* seedlings and soil.

Conflicts of Interest: The authors declare no conflicts of interest.

References

1. Ruiz, D.; San Miguel, G.; Corona, B.; Gaitero, A.; Domínguez, A. Environmental and Economic Analysis of Power Generation in a Thermophilic Biogas Plant. *Sci. Total Environ.* **2018**, *633*, 1418–1428. [CrossRef]
2. Mazzurco Miritana, V.; Patrolecco, L.; Barra Caracciolo, A.; Visca, A.; Piccinini, F.; Signorini, A.; Rosa, S.; Grenni, P.; Garbini, G.L.; Spataro, F.; et al. Effects of Ciprofloxacin Alone or in Mixture with Sulfamethoxazole on the Efficiency of Anaerobic Digestion and Its Microbial Community. *Antibiotics* **2022**, *11*, 1111. [CrossRef] [PubMed]

3. Podjanapon, P.; Jamaree, T.; Sakhonwasee, S. Optimization of Liquid Digestate Fertigation and Light Intensity for Lettuce Cultivation in Closed Plant Production System. *Environ. Control Biol.* **2023**, *61*, 9–16. [CrossRef]
4. Panuccio, M.R.; Mallamaci, C.; Attinà, E.; Muscolo, A. Using Digestate as Fertilizer for a Sustainable Tomato Cultivation. *Sustainability* **2021**, *13*, 1574. [CrossRef]
5. Mortola, N.; Romaniuk, R.; Cosentino, V.; Eiza, M.; Carfagno, P.; Rizzo, P.; Bres, P.; Riera, N.; Roba, M.; Butti, M.; et al. Potential Use of a Poultry Manure Digestate as a Biofertiliser: Evaluation of Soil Properties and *Lactuca sativa* Growth. *Pedosphere* **2019**, *29*, 60–69. [CrossRef]
6. Kwoczynski, Z.; Burdová, H.; Al Souki, K.S.; Čmelík, J. Extracted Rapeseed Meal Biochar Combined with Digestate as a Soil Amendment: Effect on Lettuce (*Lactuca sativa* L.) Biomass Yield and Concentration of Bioavailable Element Fraction in the Soil. *Sci. Hortic.* **2024**, *329*, 113041. [CrossRef]
7. Jamison, J.; Khanal, S.K.; Nguyen, N.H.; Deenik, J.L. Assessing the Effects of Digestates and Combinations of Digestates and Fertilizer on Yield and Nutrient Use of *Brassica juncea* (Kai Choy). *Agronomy* **2021**, *11*, 509. [CrossRef]
8. Panagos, P.; Ballabio, C.; Lugato, E.; Jones, A.; Borrelli, P.; Scarpa, S.; Orgiazzi, A.; Montanarella, L. Potential Sources of Anthropogenic Copper Inputs to European Agricultural Soils. *Sustainability* **2018**, *10*, 2380. [CrossRef]
9. Fagnano, M.; Agrelli, D.; Pascale, A.; Adamo, P.; Fiorentino, N.; Rocco, C.; Pepe, O.; Ventorino, V. Copper Accumulation in Agricultural Soils: Risks for the Food Chain and Soil Microbial Populations. *Sci. Total Environ.* **2020**, *734*, 139434. [CrossRef]
10. Chen, G.; Li, J.; Han, H.; Du, R.; Wang, X. Physiological and Molecular Mechanisms of Plant Responses to Copper Stress. *Int. J. Mol. Sci.* **2022**, *23*, 12950. [CrossRef]
11. Bankaji, I.; Caçador, I.; Sleimi, N. Physiological and Biochemical Responses of Suaeda Fruticosa to Cadmium and Copper Stresses: Growth, Nutrient Uptake, Antioxidant Enzymes, Phytochelatin, and Glutathione Levels. *Environ. Sci. Pollut. R.* **2015**, *22*, 13058–13069. [CrossRef]
12. Zvezdanovic, J.; Markovic, D.; Nikolic, G. Different Possibilities for the Formation of Complexes of Copper and Zinc with Chlorophyll inside Photosynthetic Organelles: Chloroplasts and Thylakoids. *J. Serb. Chem. Soc.* **2007**, *72*, 1053–1062. [CrossRef]
13. Mir, A.R.; Pichtel, J.; Hayat, S. Copper: Uptake, Toxicity and Tolerance in Plants and Management of Cu-Contaminated Soil. *BioMetals* **2021**, *34*, 737–759. [CrossRef]
14. Fedeli, R.; Celletti, S.; Loppi, S.; Vannini, A. Comparison of the Effect of Solid and Liquid Digestate on the Growth of Lettuce (*Lactuca sativa* L.) Plants. *Agronomy* **2023**, *13*, 782. [CrossRef]
15. Faran, M.; Nadeem, M.; Manful, C.F.; Galagedara, L.; Thomas, R.H.; Cheema, M. Agronomic Performance and Phytochemical Profile of Lettuce Grown in Anaerobic Dairy Digestate. *Agronomy* **2023**, *13*, 182. [CrossRef]
16. Peijnenburg, W.; Baerselman, R.; de Groot, A.; Jager, T.; Leenders, D.; Posthuma, L.; Van Veen, R. Quantification of Metal Bioavailability for Lettuce (*Lactuca sativa* L.) in Field Soils. *Arch. Environ. Contam. Toxicol.* **2000**, *39*, 420–430. [CrossRef] [PubMed]
17. Barra Caracciolo, A.; Grenni, P.; Garbini, G.L.; Rolando, L.; Campanale, C.; Aimola, G.; Fernandez-Lopez, M.; Fernandez-Gonzalez, A.J.; Villadas, P.J.; Ancona, V. Characterization of the Belowground Microbial Community in a Poplar-Phytoremediation Strategy of a Multi-Contaminated Soil. *Front. Microbiol.* **2020**, *11*, 2073. [CrossRef] [PubMed]
18. Huang, F.; Guo, Z.; Xu, Z. Determined Methods of Chlorophyll from Lemma Paucicostata. *Exp. Technol. Manag.* **2007**, *24*, e31.
19. Lichtenthaler, H.K. Chlorophylls and Carotenoids: Pigments of Photosynthetic Biomembranes. In *Methods in Enzymology*; Academic Press: Cambridge, MA, USA, 1987; pp. 350–382.
20. Strasser, R.J.; Tsimilli-Michael, M.; Srivastava, A. Analysis of the Chlorophyll a Fluorescence Transient. In *Advances in Photosynthesis and Respiration*; Papageorgiou, G.C., Govindjee, Eds.; Springer: Dordrecht, The Netherlands, 2004; Volume 19, pp. 321–362.
21. Hodges, D.M.; DeLong, J.M.; Forney, C.F.; Prange, R.K. Improving the Thiobarbituric Acid-Reactive-Substances Assay for Estimating Lipid Peroxidation in Plant Tissues Containing Anthocyanin and Other Interfering Compounds. *Planta* **1999**, *207*, 604–611.
22. Bernabé-Antonio, A.; Sánchez-Sánchez, A.; Romero-Estrada, A.; Meza-Contreras, J.C.; Silva-Guzmán, J.A.; Fuentes-Talavera, F.J.; Hurtado-Díaz, I.; Alvarez, L.; Cruz-Sosa, F. Establishment of a Cell Suspension Culture of *Eysenhardtia platycarpa*: Phytochemical Screening of Extracts and Evaluation of Antifungal Activity. *Plants* **2021**, *10*, 414. [CrossRef]
23. Huang, W.-L.; Chang, W.-H.; Cheng, S.-F.; Li, H.-Y.; Chen, H.-L. Potential Risk of Consuming Vegetables Planted in Soil with Copper and Cadmium and the Influence on Vegetable Antioxidant Activity. *Appl. Sci.* **2021**, *11*, 3761. [CrossRef]
24. Wilkins, D.A. The measurement of tolerance to edaphic factors by means of root growth. *New Phytol.* **1978**, *80*, 623–633. [CrossRef]
25. Barra Caracciolo, A.; Giuliano, G.; Grenni, P.; Cremisini, C.; Ciccoli, R.; Ubaldi, C. Effect of Urea on Degradation of Terbuthylazine in Soil. *Environ. Toxicol. Chem.* **2005**, *24*, 1035–1040. [CrossRef] [PubMed]
26. Barra Caracciolo, A.; Bustamante, M.A.; Nogues, I.; Di Lenola, M.; Luprano, M.L.; Grenni, P. Changes in Microbial Community Structure and Functioning of a Semiarid Soil Due to the Use of Anaerobic Digestate Derived Composts and Rosemary Plants. *Geoderma* **2015**, *245–246*, 89–97. [CrossRef]
27. Grenni, P.; Rodríguez-Cruz, M.S.; Herrero-Hernández, E.; Marín-Benito, J.M.; Sánchez-Martín, M.J.; Barra Caracciolo, A. Effects of Wood Amendments on the Degradation of Terbuthylazine and on Soil Microbial Community Activity in a Clay Loam Soil. *Water Air Soil. Pollut.* **2012**, *223*, 5401–5412. [CrossRef]
28. Vaish, B.; Srivastava, V.; Singh, U.K.; Gupta, S.K.; Chauhan, P.S.; Kothari, R.; Singh, R.P. Explicating the Fertilizer Potential of Anaerobic Digestate: Effect on Soil Nutrient Profile and Growth of *Solanum melongena* L. *Environ. Technol. Innov.* **2022**, *27*, 102471. [CrossRef]

29. Coelho, J.J.; Hennessy, A.; Casey, I.; Woodcock, T.; Kennedy, N. Responses of Ryegrass, White Clover, Soil Plant Primary Macronutrients and Microbial Abundance to Application of Anaerobic Digestates, Cattle Slurry and Inorganic N-Fertiliser. *Appl. Soil. Ecol.* **2019**, *144*, 112–122. [CrossRef]
30. Cristina, G.; Camelin, E.; Tommasi, T.; Fino, D.; Pugliese, M. Anaerobic Digestates from Sewage Sludge Used as Fertilizer on a Poor Alkaline Sandy Soil and on a Peat Substrate: Effects on Tomato Plants Growth and on Soil Properties. *J. Environ. Manag.* **2020**, *269*, 110767. [CrossRef]
31. Fernández-Rodríguez, M.J.; Mancilla-Leytón, J.M.; Jiménez-Rodríguez, A.; Borja, R.; Rincón, B. Reuse of the Digestate Obtained from the Biomethanization of Olive Mill Solid Waste (OMSW) as Soil Amendment or Fertilizer for the Cultivation of Forage Grass (*Lolium rigidum* Var. Wimmera). *Sci. Total Environ.* **2021**, *792*, 148465. [CrossRef]
32. Gharbi, F.; Rejeb, S.; Ghorbal, M.H.; Morel, J.L. Plant Response to Copper Toxicity as Affected by Plant Species and Soil Type. *J. Plant Nutr.* **2005**, *28*, 379–392. [CrossRef]
33. Shams, M.; Ekinci, M.; Turan, M.; Dursun, A.; Kul, R.; Yildirim, E. Growth, Nutrient Uptake and Enzyme Activity Response of Lettuce (*Lactuca sativa* L.) to Excess Copper. *Environ. Sustain.* **2019**, 2, 67–73. [CrossRef]
34. Jin, K.; Ran, Y.; Alengebawy, A.; Yang, G.; Jia, S.; Ai, P. Agro-Environmental Sustainability of Using Digestate Fertilizer for Solanaceous and Leafy Vegetables Cultivation: Insights on Fertilizer Efficiency and Risk Assessment. *J. Environ. Manag.* **2022**, *320*, 115895. [CrossRef] [PubMed]
35. Wu, W.; Lin, Z.; Zhu, X.; Li, G.; Zhang, W.; Chen, Y.; Ren, L.; Luo, S.; Lin, H.; Zhou, H.; et al. Improved Tomato Yield and Quality by Altering Soil Physicochemical Properties and Nitrification Processes in the Combined Use of Organic-Inorganic Fertilizers. *Eur. J. Soil. Biol.* **2022**, *109*, 103384. [CrossRef]
36. Slamet, W.; Purbajanti, E.D.; Darmawati, A.; Fuskhah, E. Leaf Area Index, Chlorophyll, Photosynthesis Rate of Lettuce (*Lactuca sativa* L.) under N-Organic Fertilizer. *Indian. J. Agric. Res.* **2017**, *51*, 365–369. [CrossRef]
37. Li, J.; Qiu, Y.; Zhao, Q.; Chen, D.; Wu, Z.; Peng, A.; Niazi, N.K.; Trakal, L.; Sakrabani, R.; Gao, B.; et al. Lead and Copper-Induced Hormetic Effect and Toxicity Mechanisms in Lettuce (*Lactuca sativa* L.) Grown in a Contaminated Soil. *Sci. Total Environ.* **2020**, *741*, 140440. [CrossRef] [PubMed]
38. Tiong, Y.W.; Sharma, P.; Xu, S.; Bu, J.; An, S.; Foo, J.B.L.; Wee, B.K.; Wang, Y.; Lee, J.T.E.; Zhang, J.; et al. Enhancing Sustainable Crop Cultivation: The Impact of Renewable Soil Amendments and Digestate Fertilizer on Crop Growth and Nutrient Composition. *Environ. Pollut.* **2024**, *342*, 123132. [CrossRef] [PubMed]
39. Pranckietienė, I.; Navickas, K.; Venslauskas, K.; Jodaugienė, D.; Buivydas, E.; Žalys, B.; Vagusevičienė, I. The Effect of Digestate from Liquid Cow Manure on Spring Wheat Chlorophyll Content, Soil Properties, and Risk of Leaching. *Agronomy* **2023**, *13*, 626. [CrossRef]
40. Maxwell, K.; Johnson, G.N. Chlorophyll Fluorescence—A Practical Guide. *J. Exp. Bot.* **2000**, *51*, 659–668. [CrossRef]
41. Pietrini, F.; Carnevale, M.; Beni, C.; Zacchini, M.; Gallucci, F.; Santangelo, E. Effect of Different Copper Levels on Growth and Morpho-Physiological Parameters in Giant Reed (*Arundo donax* L.) in Semi-Hydroponic Mesocosm Experiment. *Water* **2019**, *11*, 1837. [CrossRef]
42. Baker, N.R. Chlorophyll Fluorescence: A Probe of Photosynthesis In Vivo. *Annu. Rev. Plant Biol.* **2008**, *59*, 89–113. [CrossRef]
43. Bussotti, F.; Desotgiu, R.; Cascio, C.; Pollastrini, M.; Gravano, E.; Gerosa, G.; Marzuoli, R.; Nali, C.; Lorenzini, G.; Salvatori, E.; et al. Ozone Stress in Woody Plants Assessed with Chlorophyll a Fluorescence. A Critical Reassessment of Existing Data. *Environ. Exp. Bot.* **2011**, *73*, 19–30. [CrossRef]
44. Gilmore, A.M.; Hazlett, T.L.; Debrunner, P.G. Govindjee Comparative Time-Resolved Photosystem II Chlorophyll *a* Fluorescence Analyses Reveal Distinctive Differences between Photoinhibitory Reaction Center Damage and Xanthophyll Cycle-Dependent Energy Dissipation. *Photochem. Photobiol.* **1996**, *64*, 552–563. [CrossRef] [PubMed]
45. Singh, H.; Kumar, D.; Soni, V. Performance of Chlorophyll a Fluorescence Parameters in Lemna Minor under Heavy Metal Stress Induced by Various Concentration of Copper. *Sci. Rep.* **2022**, *12*, 10620. [CrossRef] [PubMed]
46. Grieco, M.; Suorsa, M.; Jajoo, A.; Tikkanen, M.; Aro, E.-M. Light-Harvesting II Antenna Trimers Connect Energetically the Entire Photosynthetic Machinery—Including Both Photosystems II and I. *Biochim. Biophys. Acta (BBA) Bioenerg.* **2015**, *1847*, 607–619. [CrossRef] [PubMed]
47. Xu, C.; Tian, Y.; Sun, Y.; Dong, L. Effects of Biogas Slurry Irrigation on Growth, Photosynthesis, and Nutrient Status of *Perilla frutescens* Seedlings. *Commun. Soil. Sci. Plant Anal.* **2013**, *44*, 3381–3390. [CrossRef]
48. Malav, L.C.; Khan, S.A.; Gupta, N. Impacts of Biogas Slurry Application on Soil Environment, Yield and Nutritional Quality of Baby Corn. *Soc. Plant Res.* **2015**, *74*, 194. [CrossRef]
49. Zheng, X.; Fan, J.; Cui, J.; Wang, Y.; Zhou, J.; Ye, M.; Sun, M. Effects of Biogas Slurry Application on Peanut Yield, Soil Nutrients, Carbon Storage, and Microbial Activity in an Ultisol Soil in Southern China. *J. Soils Sediments* **2016**, *16*, 449–460. [CrossRef]
50. Singh, R.P.; Singh, P.; Ibrahim, M.H.; Hashim, R. Land Application of Sewage Sludge: Physicochemical and Microbial Response. *Rev. Environ. Contam. Toxicol.* **2011**, *214*, 41–61.
51. Tan, F.; Zhu, Q.; Guo, X.; He, L. Effects of Digestate on Biomass of a Selected Energy Crop and Soil Properties. *J. Sci. Food Agric.* **2021**, *101*, 927–936. [CrossRef]
52. Qi, X.; Zhang, S.; Wang, Y.; Wang, R. Advantages of the Integrated Pig-Biogas-Vegetable Greenhouse System in North China. *Ecol. Eng.* **2005**, *24*, 175–183. [CrossRef]
53. Morales, M.; Munné-Bosch, S. Malondialdehyde: Facts and Artifacts. *Plant Physiol.* **2019**, *180*, 1246–1250. [CrossRef] [PubMed]

54. Liang, X.; Qian, R.; Wang, D.; Liu, L.; Sun, C.; Lin, X. Lipid-Derived Aldehydes: New Key Mediators of Plant Growth and Stress Responses. *Biology* **2022**, *11*, 1590. [CrossRef] [PubMed]
55. Tagnon, M.D.; Simeon, K.O. Aldehyde Dehydrogenases May Modulate Signaling by Lipid Peroxidation-Derived Bioactive Aldehydes. *Plant Signal Behav.* **2017**, *12*, e1387707. [CrossRef] [PubMed]
56. Nicoletto, C.; Santagata, S.; Zanin, G.; Sambo, P. Effect of the Anaerobic Digestion Residues Use on Lettuce Yield and Quality. *Sci. Hortic.* **2014**, *180*, 207–213. [CrossRef]
57. Panuccio, M.R.; Papalia, T.; Attinà, E.; Giuffrè, A.; Muscolo, A. Use of Digestate as an Alternative to Mineral Fertilizer: Effects on Growth and Crop Quality. *Arch. Agron. Soil. Sci.* **2019**, *65*, 700–711. [CrossRef]
58. Li, F.; Yuan, Y.; Gong, P.; Imazumi, Y.; Na, R.; Shimizu, N. Comparative Effects of Mineral Fertilizer and Digestate on Growth, Antioxidant System, and Physiology of Lettuce under Salt Stress. *Hortic. Environ. Biotechnol.* **2023**, *64*, 379–391. [CrossRef]
59. Sipter, E.; Auerbach, R.; Gruiz, K.; Mathe-Gaspar, G. Change of Bioaccumulation of Toxic Metals in Vegetables. *Commun. Soil. Sci. Plant Anal.* **2009**, *40*, 285–293. [CrossRef]
60. Chaturvedi, R.; Favas, P.J.C.; Pratas, J.; Varun, M.; Paul, M.S. Metal(Loid) Induced Toxicity and Defense Mechanisms in *Spinacia oleracea* L.: Ecological Hazard and Prospects for Phytoremediation. *Ecotoxicol. Environ. Saf.* **2019**, *183*, 109570. [CrossRef]
61. Kikis, C.; Thalassinos, G.; Antoniadis, V. Soil Phytomining: Recent Developments—A Review. *Soil. Syst.* **2024**, *8*, 8. [CrossRef]
62. Christiansen, K.S.; Borggaard, O.K.; Holm, P.E.; Vijver, M.G.; Hauschild, M.Z.; Peijnenburg, W.J.G.M. Experimental Determinations of Soil Copper Toxicity to Lettuce (*Lactuca sativa*) Growth in Highly Different Copper Spiked and Aged Soils. *Environ. Sci. Pollut. R* **2015**, *22*, 5283–5292. [CrossRef]
63. Audet, P.; Charest, C. Heavy Metal Phytoremediation from a Meta-Analytical Perspective. *Environ. Pollut.* **2007**, *147*, 231–237. [CrossRef] [PubMed]
64. Wolf, M.; Baretta, D.; Becegato, V.A.; Almeida, V.d.C.; Paulino, A.T. Copper/Zinc Bioaccumulation and the Effect of Phytotoxicity on the Growth of Lettuce (*Lactuca sativa* L.) in Non-Contaminated, Metal-Contaminated and Swine Manure-Enriched Soils. *Water Air Soil. Pollut.* **2017**, *228*, 152. [CrossRef]
65. Niazi, P.; Monib, A. Function of Macronutrients in Plant Growth and Human. *IJSDR Res. J.* **2023**, *8*, 1265.
66. Gomes, D.G.; Pieretti, J.C.; Rolim, W.R.; Seabra, A.B.; Oliveira, H.C. Advances in Nano-Based Delivery Systems of Micronutrients for a Greener Agriculture. In *Advances in Nano-Fertilizers and Nano-Pesticides in Agriculture*; Elsevier: Amsterdam, The Netherlands, 2021; pp. 111–143.
67. *Codex Stan 193-1995*; General Standard for Contaminants and Toxins in Food and Feed. Codex Allimentarius: Rome, Italy, 2014. Available online: https://www.fao.org/fileadmin/user_upload/livestockgov/documents/1_CXS_193e.pdf (accessed on 28 April 2024).
68. Olszyk, D.M.; Shiroyama, T.; Novak, J.M.; Cantrell, K.B.; Sigua, G.; Watts, D.W.; Johnson, M.G. Biochar Affects Essential Nutrients of Carrot Taproots and Lettuce Leaves. *HortScience* **2020**, *55*, 261–271. [CrossRef] [PubMed]
69. Ntinas, G.K.; Bantis, F.; Koukounaras, A.; Kougias, P.G. Exploitation of Liquid Digestate as the Sole Nutrient Source for Floating Hydroponic Cultivation of Baby Lettuce (*Lactuca sativa*) in Greenhouses. *Energies* **2021**, *14*, 7199. [CrossRef]
70. Wu, Y.; Wen, Q.; Chen, Z.; Fu, Q.; Bao, H. Response of Antibiotic Resistance to the Co-Exposure of Sulfamethoxazole and Copper during Swine Manure Composting. *Sci. Total Environ.* **2022**, *805*, 150086. [CrossRef]
71. Lwin, C.S.; Seo, B.-H.; Kim, H.-U.; Owens, G.; Kim, K.-R. Application of Soil Amendments to Contaminated Soils for Heavy Metal Immobilization and Improved Soil Quality—A Critical Review. *Soil. Sci. Plant Nutr.* **2018**, *64*, 156–167. [CrossRef]
72. Järvan, M.; Edesi, L.; Adamson, A.; Võsa, T. Soil Microbial Communities and Dehydrogenase Activity Depending on Farming Systems. *Plant Soil. Environ.* **2014**, *60*, 459–463. [CrossRef]
73. Cycoń, M.; Mrozik, A.; Piotrowska-Seget, Z. Antibiotics in the Soil Environment—Degradation and Their Impact on Microbial Activity and Diversity. *Front. Microbiol.* **2019**, *10*, 338. [CrossRef]
74. Moore, F.; González, M.-E.; Khan, N.; Curaqueo, G.; Sanchez-Monedero, M.; Rilling, J.; Morales, E.; Panichini, M.; Mutis, A.; Jorquera, M.; et al. Copper Immobilization by Biochar and Microbial Community Abundance in Metal-Contaminated Soils. *Sci. Total Environ.* **2018**, *616–617*, 960–969. [CrossRef]

environments

MDPI

Article

Combination between Composting and Vermicomposting of OFMSW: A Sicilian Case Study

Enrico Licitra *, Maria Gabriella Giustra, Gaetano Di Bella and Antonio Messineo

Department of Engineering and Architecture, University of Enna "Kore", Cittadella Universitaria, 94100 Enna, Italy; mariagabriella.giustra@unikore.it (M.G.G.); gaetano.dibella@unikore.it (G.D.B.); antonio.messineo@unikore.it (A.M.)

* Correspondence: enrico.licitra@unikore.it; Tel.: +39-3315902276

Abstract: In nature, earthworms process enormous quantities of plant debris, ingesting and converting them into vermicast, the final product of their digestion process. Vermicast is widely recognized as an organic fertilizer suitable for plants, usually obtained from the transformation of animal manure. Indeed, controlled vermicomposting of the organic fraction from municipal solid waste (OFMSW) has yielded contradictory results, limiting the extension of laboratory results to larger-scale initiatives. This study aims to analyze a combined composting–vermicomposting process using *Eisenia fetida* (also known as Californian red worm) for the treatment of OFMSW (containing different impurities, such as heavy metals), mixed with a suitable substrate for earthworms, consisting of a blend of animal bedding and pre-selected food scraps (SOM—pre-selected organic material). Different fractions from a municipal composting plant were tested for the biological process: raw OFMSW, pre-composted (PC, organic matrix that has completed the thermophilic biodegradation phase), and compost (C). Laboratory tests involved five different applications with varying mixing proportions, partly aided by the addition of OM. The physical–chemical parameters (e.g., pH, temperature, moisture) and worm growth rate in the different treatments were measured and compared. The results showed that the process improved the quality of the final product, especially for the selected matrices, and revealed a significant reduction in the carbon-to-nitrogen ratio (lower than 14 in all tests) when vermicomposting was applied to the mixed matrices and SOM. Worms increased during the process in net weight and growth rate, even if there was an accumulation of heavy metals in the "worm tea". Worm mass reached a little over double the initial value in all tests (from 200 g to more than 500 g in the SOM test), except for an increase of only 87% in the OFMSW test, while heavy metal content in the solid matrix was reduced in all tests compared to the starting content.

Keywords: vermicomposting; *Eisenia fetida*; heavy metals; OFMSW; nutrients; compost

1. Introduction

The production of solid waste (SW) inevitably tends to increase with population growth and economic and social development [1–3]. Solid waste includes municipal (MSW), industrial, agricultural waste, and sewage sludge; there is a projected progressive increase in the production of such waste [4]. Unfortunately, conventional waste disposal methods (such as landfills and incinerators) have more significant environmental impacts compared to treatments based on biological processes. These biological processes, especially those focusing on material recovery strategies, are generally more efficient than thermal transformation processes. Specifically, treatment methods based on biotechnological processes offer many advantages, including high speed, ease of control, low costs, high environmental and health acceptability, and potential applicability at different scales [5].

Composting is the most widely used and studied biological process for managing the organic fraction of MSW [6]. On the other hand, vermicomposting is inspired by a similar biological (aerobic) process operated by the combined activity of worms and bacteria, rather than merely aerobic bacteria [7].

Composting is a process of aerobic biodegradation of organic waste carried out by microorganisms in which organic waste is converted into mineral compounds such as CO_2, H_2O, and NH_4, as well as organic compounds rich in humus (a complex of essential organic substances for plant nutrition that result from the decomposition; it is strongly adsorbent, and it has a blackish brown color). These humus-rich compounds are suitable for improving the physical properties of the soil [8,9]. On the other hand, vermicomposting is a biological process similar to the composting process, but it is not exclusively bacterial, and nutrients contained in organic waste are transformed into a stable, available, and nutrient-rich product for plant growth [10].

In vermicomposting, the oxidation and biological stabilization of organic waste depend on the interaction between worms and microorganisms in the mesophilic transformation process. Specifically, the activity of worms increases and enhances the organic matrix intended for microbial consumption [11–13]. The final vermicompost is a stabilized and homogeneous material with high nutrient content and low toxicity, offering several advantages, such as excellent cation exchange capacity and a higher nutrient retention capacity [14,15].

Worms are the animals that process the most substantial amounts of soil in nature [16]. They consume dead plant debris, animal excrement, and soil, thereby contributing to positive soil transformations and promoting the mineralization of dead plant matter [17]. However, despite the variety and abundance of worms in nature, the controlled use of worms for treating organic waste has so far been largely limited to the vermiculture of animal manure and, to a lesser extent, domestic food waste [18].

In the management of MSW, there are other more recalcitrant vegetable fractions such as leaf waste, weeds, crop residues, and toxic elements derived from imperfect upstream separation. For this reason, vermicomposting applied to OFMSW has shown many limitations and contradictory results, restricting its widespread application. Even if vermicomposting is a cleaner process than composting, as it requires much less energy due to lower temperatures during the process and does not require periodic turning of piles or forced aeration thanks to the activity of worms, there are doubts about long-term results and the real negative quality influences due to the presence of certain pollutants in municipal waste (even if selected) that currently confine the application of vermicomposting mainly to the transformation of waste from zootechnics, bovine manure, and selected food waste at a mono-residential level. In this context, for example, it cannot be ignored that the presence of potential toxic elements, such as heavy metals, in the waste can threaten soil quality and the worm's growth. Nevertheless, the biochemical transformations of heavy metals during the vermicomposting process are still controversial, despite the current compost standard setting limits on total concentrations of some toxic metals. Several studies have reported increases in the total concentrations of heavy metals during vermicomposting [19–21], while other studies have obtained a completely opposite behavior [22]. Probably, the different observed development patterns are closely related to site-specific bioaccumulation mechanisms and the evolution of substance reduction in terms of biological adaptation [23,24]. Therefore, the activity of worms can influence the solubility and bioavailability of heavy metals, modifying the speciation of the vermicast [25,26]. Different authors have reported that earthworms secrete a mucus-like substance under metal stress that aids in reducing the bioavailability of metals and bioaccumulating them on the surface of their tissues [27,28].

In this scenario, the scientific interest in combining composting and vermicomposting processes is highly relevant. In this optimized transformation pathway, a wide range of organic waste, including those resistant to decomposition, can potentially be decomposed without compromising process efficiency. Thus, the combination of composting with the vermicomposting process could be suitable for the biodegradation of solid waste from municipal selective collection, regardless of its quality [8,29]. Indeed, the excessive use of chemical fertilizers has generated several environmental problems that can be solved by the use of biofertilizers, such as municipal solid waste composts and vermicomposts, which are natural, beneficial, and ecologically friendly [30].

There are important factors that can affect the process of vermicomposting, among which pre-composting is included. Pre-composting is usually preferred to eliminate anaerobic conditions and remove any volatile gases potentially toxic to the earthworms [31].

Considering the large volume of OFMSW produced and the need for proper management and transformation of these waste materials into reusable and environmentally friendly products, this study aims to evaluate the combined process of composting–vermicomposting for the management of OFMSW, potentially complementing the process with the addition of preselected organic matrices. Due to legislative difficulties in the introduction of OFMSW as a suitable substrate for the vermicomposting process, the study also aimed to show the advantages of the biological process carried out by not only microorganisms but also earthworms. These advantages are not referred to only in terms of the time required but also in terms of the quality of the final product. The experimentation was based on laboratory batch tests using a pilot vermicomposting system, in which the physical–chemical parameters, worm growth rate, and their growth characteristics (e.g., accumulation of metals in tissues, production of worm tea) were monitored and measured during the transformation processes of real organic matrices derived from a composting plant placed in the province of Catania (Italy).

2. Materials and Methods

2.1. Composting Plant

The composting plant from which the matrices subject to study have been sampled is located within the territory of the Province of Catania (Sicily), covering an area of approximately 25,500 square meters.

The receptive capacity of the facility is currently about 150,000 tons per year of OFMSW, with a processing line that includes the following:

(1) Weighing and reception;
(2) Unloading and storage of lignocellulosic waste;
(3) Wood shredding and storage of shredded wood;
(4) Unloading and storage of organic waste;
(5) Waste mixing;
(6) Accelerated and intermediate bio-oxidation;
(7) Stabilization, maturation, and storage of raw compost;
(8) Final screening;
(9) Storage of finished compost.

Within the reception and preparation area, a preliminary operation of mixing different wastes with shredded wood is planned in order to improve the characteristics of the mixture in terms of composition and moisture for the development of biochemical processes and the quality of the final product. Following mixing, the material is transported into the bio-cells, where the composting process begins. The impermeable flooring is equipped with a leachate collection system. The aeration system can be activated in both the "supply" and "exhaust" modes; aeration ensures control of temperature and oxygen content. After the bio-oxidation phase, which generally lasts an average of 14 days, the material is transferred to the stabilization–maturation section, lasting at least 75 days. Once the maturation phase is completed, the product undergoes a final refinement operation, consisting of screening to remove coarse materials.

The samples subject to the experimental tests conducted at the Laboratory of Environmental Sanitary Engineering of the University of Enna involved the collection of three samples of approximately 30 kg each:

- OFMSW, collected before the mixing phase with bulking agents (shredded wood).
- PRE-COMPOST, collected after the accelerated bio-oxidation phase.
- COMPOST, stabilized fraction collected after the maturation phase before screening.

In order to facilitate a comparative experimental investigation with regular matrices, a sample of selected organic material from household waste was concurrently prepared. This sample represents the "blank" reference, free from impurities. In fact, it consisted of properly differentiated food scraps to prevent the possible presence of non-organic contaminants. Within that SOM was also included an amount of rabbit manure and vegetable waste used before the start of the tests to ensure worm acclimatization. The analytical characterization of each (initial) sample is presented in Table 1.

Table 1. Initial characterization of organic substrates.

Parameters	UM	OFMSW	PC	C	SOM
Organic content (O.C.)	$[g/kg_{dm}]$	323.52	284.87	268.47	321.76
Nitrogen content (N)	$[g/kg_{dm}]$	10.97	12.89	13.84	7.54
Phosphorus content (P)	$[g/kg_{dm}]$	0.525	0.809	0.881	0.389
O.C./N	--	29.5	22.1	19.4	42.7
N/P	--	20.9	15.9	15.7	19.4
pH	--	4.0	6.0	6.0	5.5
Moisture	%	85	85	85	95

2.2. Experimental Setup

The experimental study involved the use of "prefabricated" vermicomposters provided by the company CONITALO, a leader in the vermiculture sector since 1979 (https://www.conitalo.it/, accessed on 16 June 2024). Each composting unit consisted of a lower compartment for leachate accumulation and three additional compartments (of which only 2 were utilized for the purposes of the proposed study) for placing the composting matrices. The bio-treatment compartment (the upper than leachate accumulation compartment) essentially comprised a bedding system in which the worms could be introduced within a thin layer of organic material that would be consumed over time. The organic matrices to be tested were added to the upper platforms, which were accessible to the worms residing in the lower layers. It was important that the various layers of organic material in each compartment were not too thick (maximum size 4–5 cm) to allow the worms to colonize and degrade the entire depth.

Five tests were planned using five different composting units, studying the four matrices described in the preceding paragraph (OFMSW, PC, compost, and SOM). Additionally, a "mixed" test was conducted by combining compost and selected organic material (consisting of food scraps from household waste). This was performed to ensure additional organic support in case the matrix composted in the actual composting facility was excessively depleted of substrate.

Each vermicomposter was initiated with 3 kg of matrix/bedding for the start of the trial.

The layout of the vermicomposters and the experimental setup are illustrated in Figure 1, while Table 2 provides the percentage compositions of the matrices arranged in each compartment for the 5 tests.

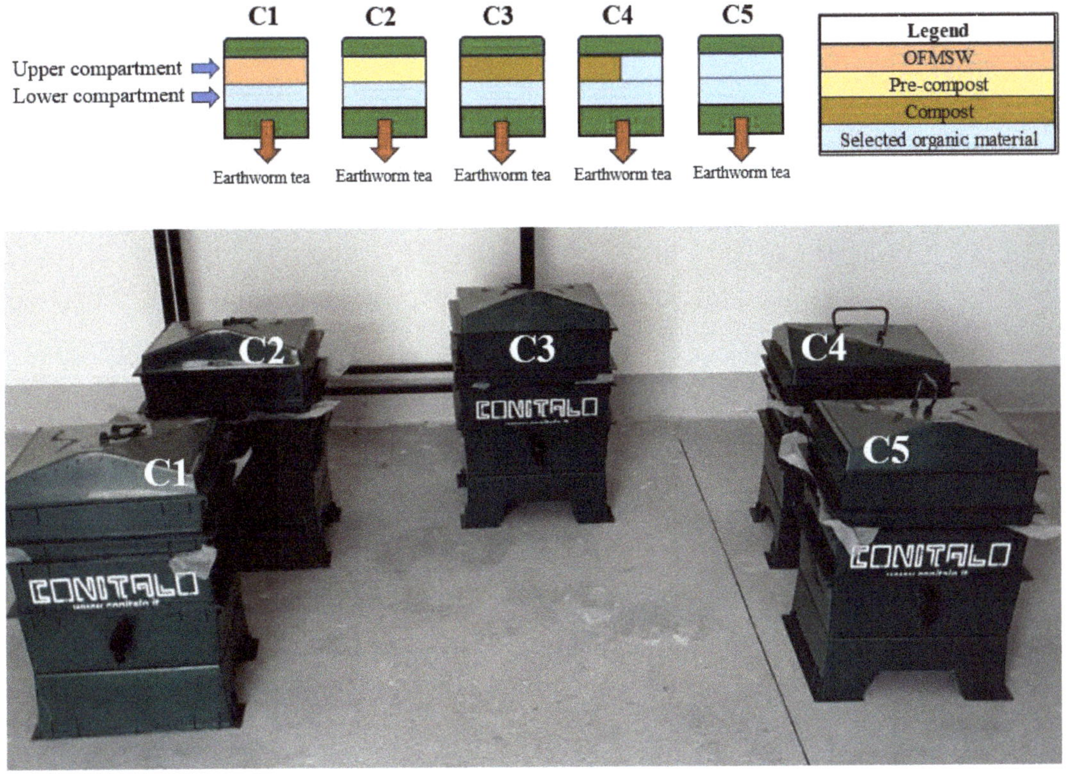

Figure 1. Test battery and details of the installed vermicomposting bins.

Table 2. Percentage composition of the different organic substrates in vermicomposters.

Vermicomposters Layout	TEST	OFMSW	PC	C	SOM
Upper compartment	C1	100%	-	-	-
	C2	-	100%	-	-
	C3	-	-	100%	-
	C4	-	-	50%	50%
	C5	-	-	-	100%
Lower compartment	C1	-	-	-	100%
	C2	-	-	-	100%
	C3	-	-	-	100%
	C4	-	-	-	100%
	C5	-	-	-	100%

2.3. Analytical Methods

Measurements of humidity, pH, and temperature were carried out using insertion probes directly on the solid matrix. The nitrogen content was evaluated using a standardized method that involved an initial preparation phase, including the grinding of the organic matrices to be analyzed and a preliminary assessment of moisture (in an oven for 24 h at 105 °C). Starting from the initial moisture value, the necessary aliquot of the ground matrix was determined to be mixed with deionized water, ensuring the correct dry matter-to-water ratio (0.5 g of dry matter and 500 mL of water or 1 g DM with 1 L of water). The subsequent phase included initial mechanical agitation aimed at breaking up larger-sized granules and creating a homogeneous suspension, followed by gentle agitation of the samples (on a magnetic stirrer) for 30 min. At the end of the agitation phase, an aliquot of the slurry needed for total nitrogen analysis was collected for cuvette testing and then spectrophotometrically analyzed (using the chromotropic acid method).

The phosphorus content was evaluated using ICP (*Inductively Coupled Plasma*) spectrometry. Sample preparation involved initial grinding or pestle–mortar pulverization to render the solid matrix homogeneous and destructured. Subsequently, a volume of 6 mL of nitric acid and 3 mL of hydrochloric acid were added to 1 g of wet sample to prepare for sample digestion at 145 °C for 2 h. After filtration (0.45 microns), a 50 mL sample was subjected to spectrometric analysis.

Regarding the content of organic matter in the samples and vermicasts of each test, two measurements were conducted: organic carbon (O.C.) content and organic matter content using the Walkley-Black method [32]. This method involved the determination of O.C. by titration using initially potassium bichromate and sulphuric acid (to oxidize organic carbon) and subsequently iron sulfate heptahydrate as a titrating agent. O.C. is a measurable component of soil organic matter (O.M.). Specifically, O.C. (<O.M.) is the "available" measurable component of soil organic matter that constitutes a small portion of the soil but plays a fundamental role in its physical, chemical, and biological functions. The organic carbon content of soils varies with soil type, as well as changes in land use and, to an even greater extent, with different agricultural practices (e.g., organic fertilization in regions where livestock farming is prevalent). For this reason, C/N ratios have been referenced to the organic carbon content.

Lastly, heavy metal content was assessed using ICP spectrometry, which concerned both the analysis of solid matrices and worms. Sample preparation was the same as for the above-mentioned determination of phosphorus content.

2.4. Analysis of Earthworms' Growth

The worms, in particular the *Eisenia fetida* species [33], were provided by the same company that supplied the vermicomposters. The provided worms were young and kept in the mother culture in 200 g packages. Initially, the worm biomass was taken and acclimated in mixed cultures of rabbit manure and selected organic material before being placed in the vermicomposters.

Specifically, before the vermicomposting experiment, rabbit manure and vegetable waste were mixed in a 2:1 (w/w) ratio for the first 10 days and a 1:1 (w/w) ratio in the following 10 days. To ensure better aeration and humidity control while simultaneously removing any toxic gases, daily manual mixing of the bedding was performed. The prepared matrix was placed in the lower compartment of each vermicomposter. The worms in the lower tray were free to move to the upper tray, where, on day 0 (t = 0 days), the matrix to be investigated was allocated, as described in the previous paragraph.

The initial inoculation was carried out with 200 g of pre-acclimated worms placed in the lower compartment of each vermicomposter. The measurement of worm growth was comprehensively assessed only at the end of the experimentation (t = 180 days) when watermelon scraps were placed inside the vermicomposters in a separate compartment below the lid and specifically in the upper compartment than the one in which the organic matrices subject to testing were initially placed. This was performed daily to separate the worms attracted to the specific type of organic matter (rich in water and sugars).

On the other hand, intermediate analyses at 15 and 90 days were conducted on 5 statistically significant samples taken from the center and the 4 corners of both trays. The worms present in the samples were weighed and reported as a proportion of the total mass. Specifically, the intermediate analyses involved the sampling of 5 homogenized aliquots (50 g) from each tray.

To proceed with the physical–chemical analysis (see Section 2.3), the samples were initially air-dried at room temperature, crushed, and sieved to 0.5 mm. The worms, manually removed beforehand, were weighed again after being washed with deionized water to remove adhering material. They were then reintroduced into their specific vermicomposter.

For the assessment of metal bioaccumulation, approximately 50 g of worms from each significant compartment (C1–C5) were randomly selected and kept in the dark for 24 h. This was necessary to allow the worms to release their intestinal content, which could affect accurate measurements, especially if they had ingested a substantial amount of organic material relative to their weight. Subsequently, after washing them with ultrapure water, freeze-drying was carried out for the analysis of heavy metals (only for day 180) [34].

3. Results

3.1. Changes in pH, Temperature, Moisture, and Compostable Volume

Figure 2 shows measurements of process control parameters as well as the volume changes for each vermicomposter (C1–C5). All the reactors were maintained at room temperature within a shaded and ventilated storage area. As a result, the temperature of the compostable mass followed normal seasonal fluctuations, stabilizing around 25 ± 3 °C after sixty days.

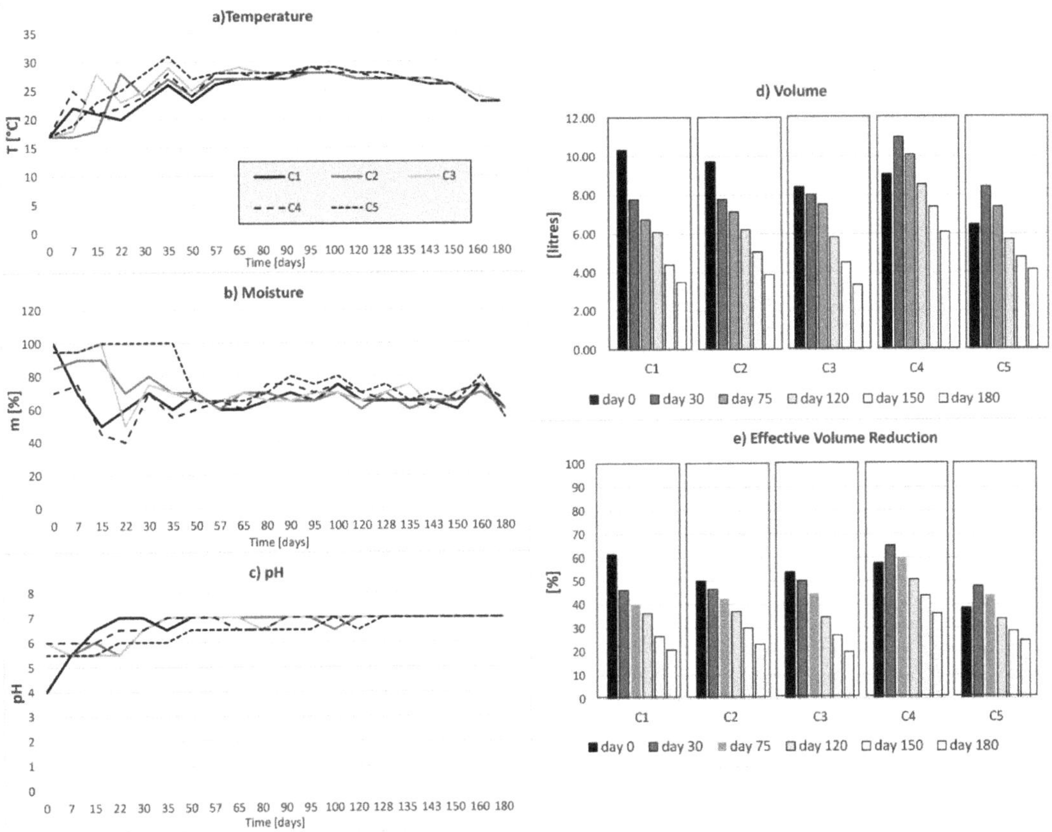

Figure 2. Variation in temperature (**a**), moisture (**b**), pH (**c**), volume (**d**), and percentage volume reduction (**e**).

Humidity was monitored through continuous and appropriate "moisturizing" interventions. Specifically, the goal was to maintain humidity around 70 ± 10% so as not to hinder the action of the worms and the necessary aerobic conditions for their development.

Only for samples C4 and C5, it was necessary to introduce an appropriate amount of coconut fibers in the initial phase to reduce the excessive initial humidity (>90%), which is characteristic of selected organic material accumulations.

Regarding pH values, it was observed that the vermicompost in each test reached an almost neutral value, although starting from different values among them. The initial values were notably acidic for the compost and selected organic material (C1 and C5) and slightly acidic for all the others.

Possible contributions to pH variations could be attributed to the differential production of CO_2, NO_3–N, and organic acids during the process [35]. In general, the increase in pH of vermicompost, compared to other composting processes, is due to greater loss of organic matter and the higher accumulation of various mineral salts in their available forms (e.g., phosphate, ammonium, potassium) in vermicompost [35,36].

As the transformation process progressed, the mass of material in the vermicomposter gradually increased in density at the expense of a significant reduction in volume.

The most substantial volume reduction was observed in test C1 concerning the OFMSW (over 65%), while for tests C2 and C3, it remained at lower values (37–40%).

The last two tests, conducted with carefully selected organic matter, recorded smaller volume reductions and higher densities. It is important to note that in this case, after the first two weeks, the volume of the material to be transformed was even greater due to the addition of coconut fibers, with an amount of about 250 g uniformly distributed in each of the two vermireactors.

3.2. Changes in Available Organic Matter and C/N Ratio

Figure 3 shows the trend of the organic component, nutrients, and the O.C./N (N/P) ratio during the experimentation for the five vermicomposting tests. As highlighted earlier, a loss of organic matter was observed. In general, the most significant loss of organic mass (O.M. and O.C.) was observed for the initially more "biodegradable" matrices, in the order C5–C1 (~8.5%) and C4 (~6.5%). The matrices related to pre-compost and compost from the actual facility, having undergone prior bacterial transformation in the composting plant, showed a lower loss of organic components (maximum 1–3%). In a complementary manner but with more significant percentage variation, the increase in total nitrogen (N) was notably recorded in the vermicompost obtained from tests C5, C4, and C1 (198.6%, 138.5%, and 128.5%, respectively) compared to tests C2 and C3 (106.3% and 70.8%, respectively).

A similar trend was observed for phosphorus (P), although due to its low initial concentration, the % increase seems more pronounced (concentration values were even 8–9 times higher than the initial ones). It is worth noting that while the increase in phosphorus follows a linear trend, nitrogen significantly increased in the second half of the experimental period once the stationary phase had stabilized.

Regarding the development of vermicast, it is evident that the reduction in organic matter is correlated with the reduction in the dry mass of the organic waste discussed in the previous paragraph. This is the result of the varying yields of decomposition and mineralization of the starting matrices and the transformations into vermicompost with different and specific concentrations of enriched nitrogen [22].

Furthermore, the carbon-to-nitrogen ratio (O.C./N ratio), where the organic component refers to the bioavailable portion related to the organic carbon, is one of the most common indicators used to estimate compost maturity and suitability for field applications [20]. Vermicomposting reduced the O.C./N ratio by 69.1% for C5, 60.8% for C4, 59.9% for C1, 52.1% for C3, and 43.5% for C2.

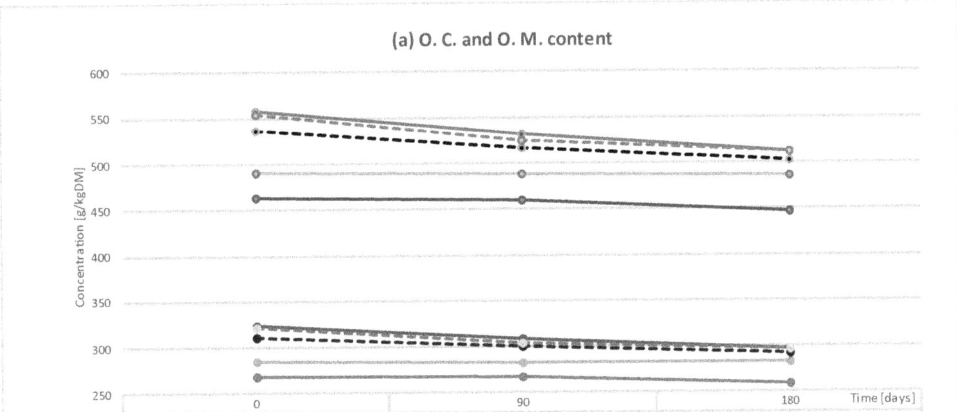

	0	90	180
O.C.-C1	323.52	308.32	296.57
O.C.-C2	284.87	282.98	282.12
O.C.-C3	268.47	266.76	258.87
O.C.-C4	311.12	299.83	291.3
O.C.-C5	321.76	304.21	296.67
O.M.-C1	557.75	531.54	511.29
O.M.-C2	491.12	487.86	486.37
O.M.-C3	462.84	459.89	446.29
O.M.-C4	536.37	516.91	502.20
O.M.-C5	554.71	524.46	511.46

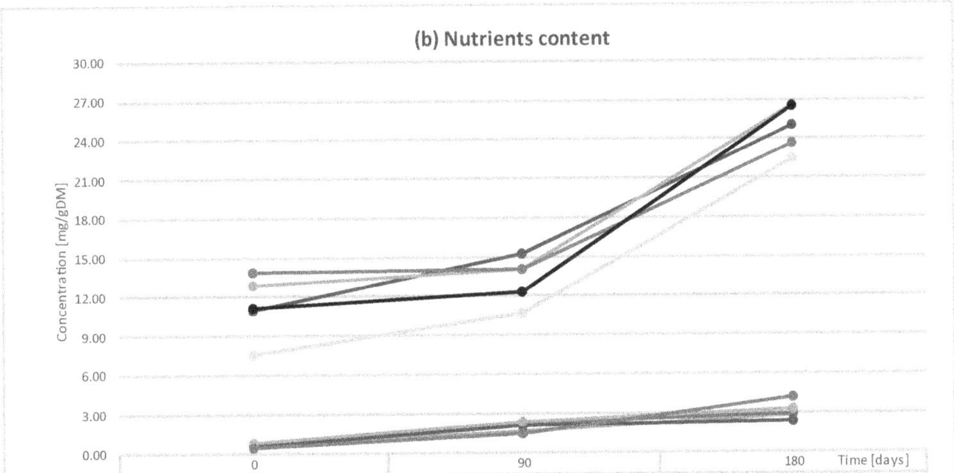

	0	90	180
N-C1	10.97	15.23	25.06
N-C2	12.89	14.02	26.60
N-C3	13.84	14.01	23.63
N-C4	11.11	12.35	26.50
N-C5	7.54	10.69	22.50
P-C1	0.525	1.60	2.82
P-C2	0.809	2.29	2.92
P-C3	0.481	2.03	2.35
P-C4	0.809	2.38	3.27
P-C5	0.389	1.40	4.21

Figure 3. *Cont.*

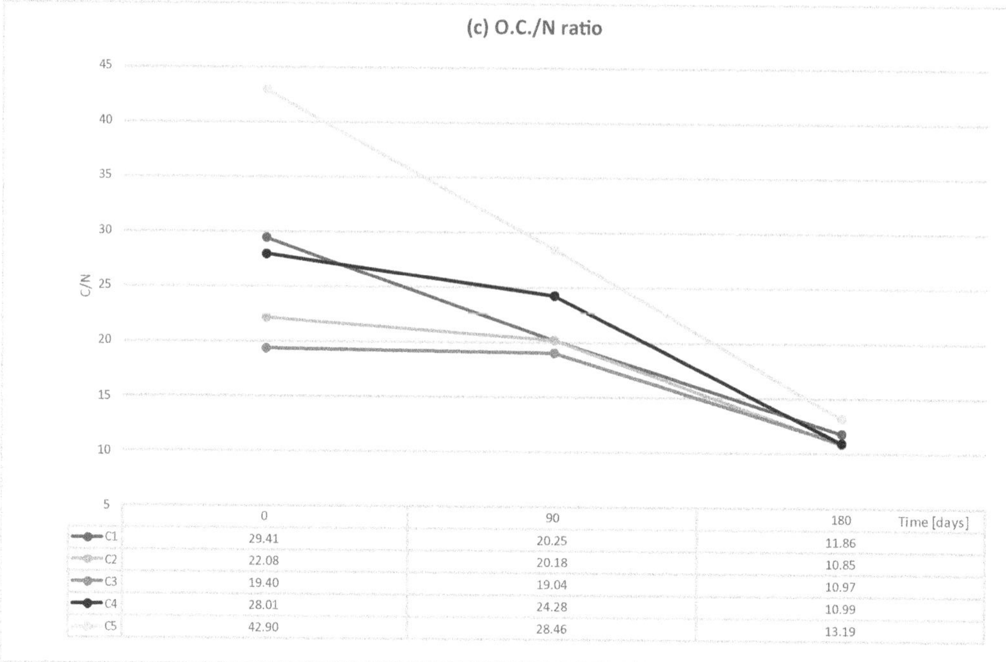

Figure 3. Evolution of organic carbon and organic matter content (**a**), nutrient content (**b**), and C/N ratio (**c**) for each test.

These values are generally higher than those observed in the literature for biological transformation through composting [34]. This confirms the simultaneous and opposite trends of O.C. and nitrogen. In absolute terms, the O.C./N ratio was found to be below 14 in all tests, with lower values observed for matrices from FORSU and the composting plant ($\leq$11). Slightly higher values were observed for matrices containing selected organic matter, still below 12 for C4 and $\leq$13.5 for C5.

In this context, it is worth noting that Suthar (2010) [20] had previously reported that a C/N ratio < 20 indicates acceptable maturity in finished compost, but a ratio of 15 is preferred for agricultural applications. In general, an O.C./N ratio below 12 indicated that the vermicompost possessed favorable properties for potential field applications.

3.3. Biomass Growth

Figure 4 illustrates the growth trend of the worm mass in the vermicomposters. At the end of the experiment, the total worm mass increased in all tests. As expected, the most significant increase was achieved for C5, where the matrix obtained from selected organic material was investigated. In this case, the worm mass nearly tripled.

For the intermediate matrices, C2, C3, and C4, the increase in worm mass reached a little over double the initial value. In the case of C1, however, the worm mass increased by only 87% (less than double).

Certainly, the growth results are directly linked to the availability of nutrients and organic matter, but above all to the initial growth environment. In this regard, intermediate measurements taken after 2 weeks highlighted a significant biomass loss due to the unsuitable conditions of the starting matrix.

Specifically, the worms "cultivated" in the lower compartment (optimal bedding), identical for all tests, found themselves transforming a very different matrix in the upper compartment. Some matrices, moreover, were very "complex" and unsuitable for the ordinary growth scenario; consequently, the majority of the worms "died".

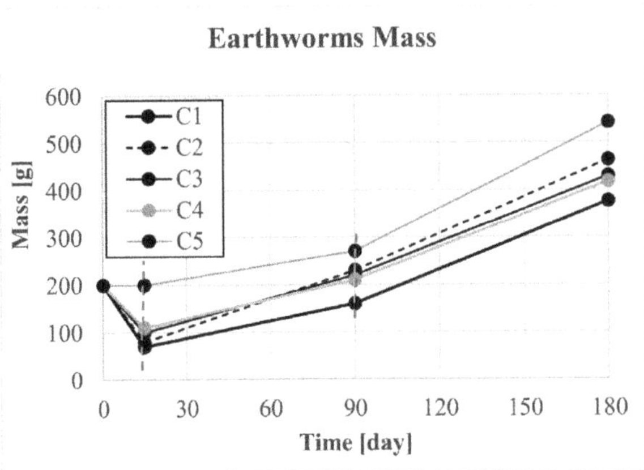

Figure 4. Growth of earthworms' mass in the five tests.

Mortality was notable for C1 and C2 (65% and 60%) and significant for C3 (50%). On the contrary, no mortality was recorded in C5. In C4, where compost and selected organic material were mixed, mortality was contained to less than 50%.

The cause of this initial "stabilization" was likely due to the particularly acidic and putrescible environment of the matrix (more pronounced in C1, as shown in Figure 3c), which was profoundly different from the degradable matrix obtained from the mixture of organic and bedding litter manure (lower compartment).

It is important to highlight that the surviving biomass rapidly adapted, generating new worm biomass adapted to the harsh growth conditions, as shown by the increasing values on the 90th day. The final result allowed defining an active matrix composed of both adult worms and worms "in the development phase" (which in some tests reached up to 50% of the total) that still contributed to the transformation of the vermicast, albeit with growth outcomes lower than those obtained in other experiments based on matrices different from municipal organic waste.

Finally, it is possible that the growth and the need for "initial adaptation" were influenced by the bioaccumulation of heavy metals in their tissues, as confirmed by the studies [23,34] or the specific physiology of the metal [37,38].

In particular, as discussed in the subsequent paragraph, earthworms can potentially accumulate a certain amount of toxic substances, such as metals, in their tissues and inevitably introduce these toxic substances into terrestrial food chains.

3.4. Heavy Metal Bioavailability

Figure 5 presents the content of four metals of interest (Cu, Zn, Ni, and Cd) in the vermicast and leachate from the five tests carried out. The selection of the investigated metals was based on the characterization of the initial matrices, which, due to their origin (urban solid waste), contained significant concentrations of some heavy metals compared to others. This circumstance was partly attributed to the significant impurities in the starting matrix (batteries, plastics and microplastics, cans, etc.).

It is important to highlight that the concentration of heavy metals in the initial matrices was higher in the untreated raw matrix (OFMSW, C1) compared to the processed matrices (pre-compost, C2, and compost, C3). Lower content was recorded in matrices containing pure selected material (C5) or combined material (C4). However, in all tests and for all the analyzed metals, the initial values were reduced during the vermicomposting process.

Following are some noteworthy observations from Figure 5. In neither the starting nor final matrices did values exceed the limits set by Legislative Decree 75/2010.

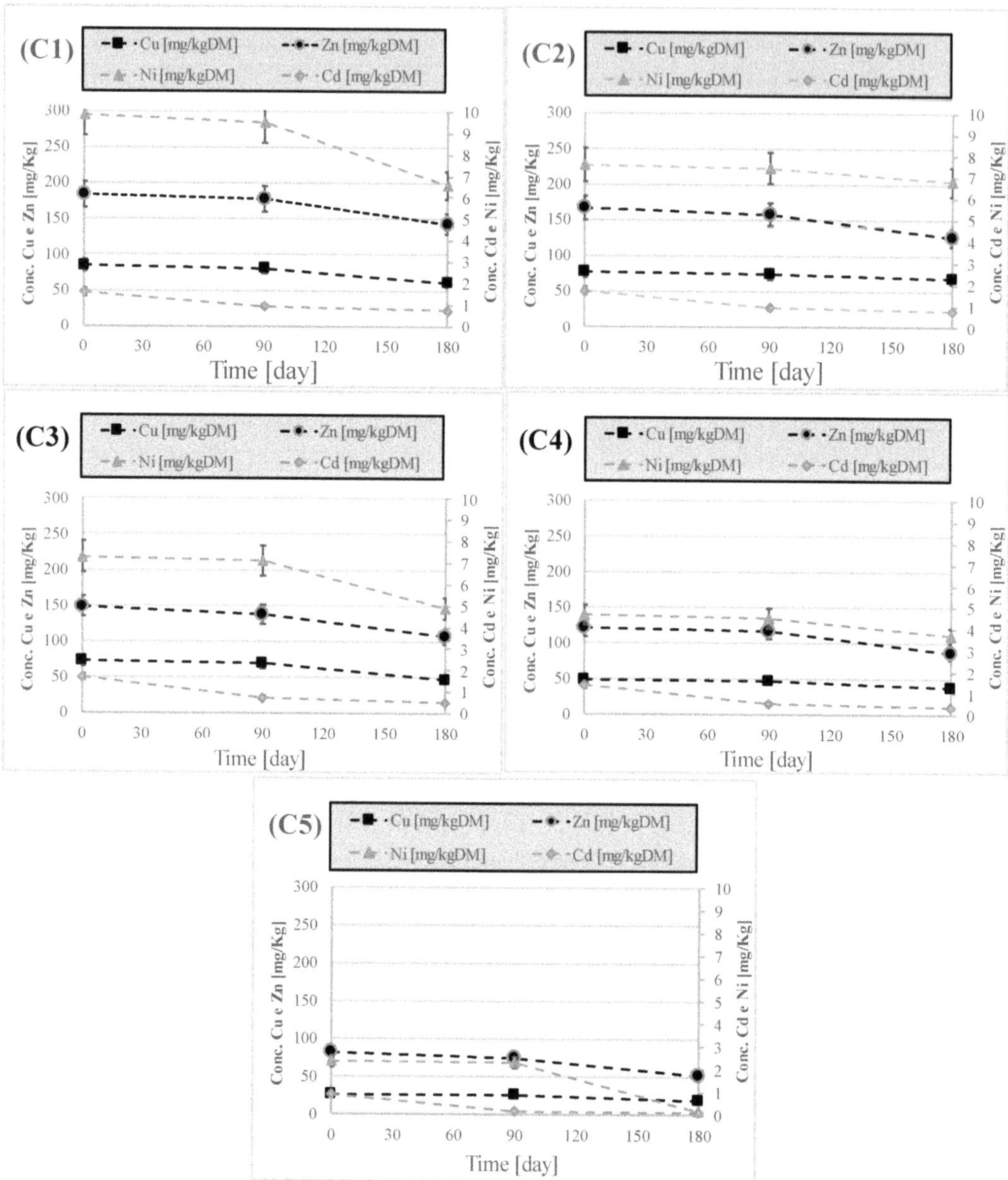

Figure 5. Heavy metal content in the solid matrix during the experimental period in each vermicomposter.

First and foremost, it should be highlighted that the bioaccumulation of metals, evaluated also in the tissue of the worms, was found to be significantly lower than that observed by other researchers [34].

Probably, this can be attributed to the fact that the worms required a significant adaptation period to the working conditions, with quite notable initial mortality rates.

Furthermore, the net growth observed at the end of the six-month observation period was lower compared to that observed in the transformation of animal manure. For this reason, a portion of the heavy metals was "concentrated" in the vermicast and eluate liquid (partially produced by the biomass) rather than in the worms themselves.

Cd has shown the most significant reduction in concentration in the final vermicast, exceeding 50% for C1 and C2, nearly 70% for C3, and surpassing 70% for C4. The reference matrix for C5, pure selected organic material, exhibited a reduction of approximately 90% in Cd. With the exception of Cd, all other heavy metals exhibited a general reduction trend mainly concentrated in the second period of the test (between days 90 and 180). As anticipated, the concentrations of heavy metals in the vermicast obtained from pure organic material (C5) were negligible.

The observations made in the solid matrix were confirmed by the concentrations of metals obtained in the liquid collected from each composting unit. Figure 6 shows the volumes and metal concentrations in the liquid samples of each test.

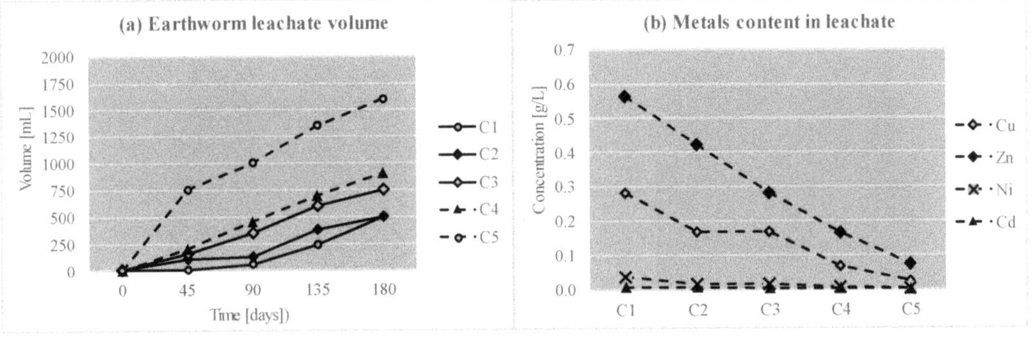

Figure 6. Volume (**a**) and metal content (**b**) in the produced leachate (earthworm tea) for each test.

It is observed that the volume of "leachate" produced by the "domestic" organic matrices was significantly higher than that collected from tests conducted with raw matrices (OFMSW in C1) and semi-processed matrices (pre-compost and compost, C2 and C3).

This is likely due to the different consistency/density of the analyzed matrices and the developed vermicast (Figure 7); in particular, this aspect influenced the starting value of FC (field capacity) and, therefore, percolation. On the other hand, the values of different initial moisture, discussed earlier, support the volumetric differences in collection.

In any case, as reported in Figure 6b, the metal concentrations are significantly high in C1, C2, and C3; they are much lower in C4 and negligible in C5. This result suggests that the potential use of "worm tea" as a fertilizer obtained from OFMSW matrices (raw or processed) must be seriously evaluated/paid attention to if the purity of the matrices is compromised by suboptimal waste separation.

On the other hand, for the same reason, in the studied case, the liquid collected in the vermicomposting trays has been referred to as "leachate" since the dominant formation process primarily involves elution processes rather than biological formation due to the earthworms' metabolism.

Conversely, the liquid collected from tests conducted with selected organic material from household environments, mixed (C4) or pure (C5), was found to be in line with the research conducted in the literature [39–42].

Finally, it should be emphasized that in this experimental work, bioaccumulation in the earthworm tissues was found to be secondary to the leaching processes described above. In general, it was observed (Figure 8) that the most bioaccumulable metal was Cd, followed by Zn. Conversely, Cu and Ni were predominantly found in the leachate rather than in the earthworm tissues.

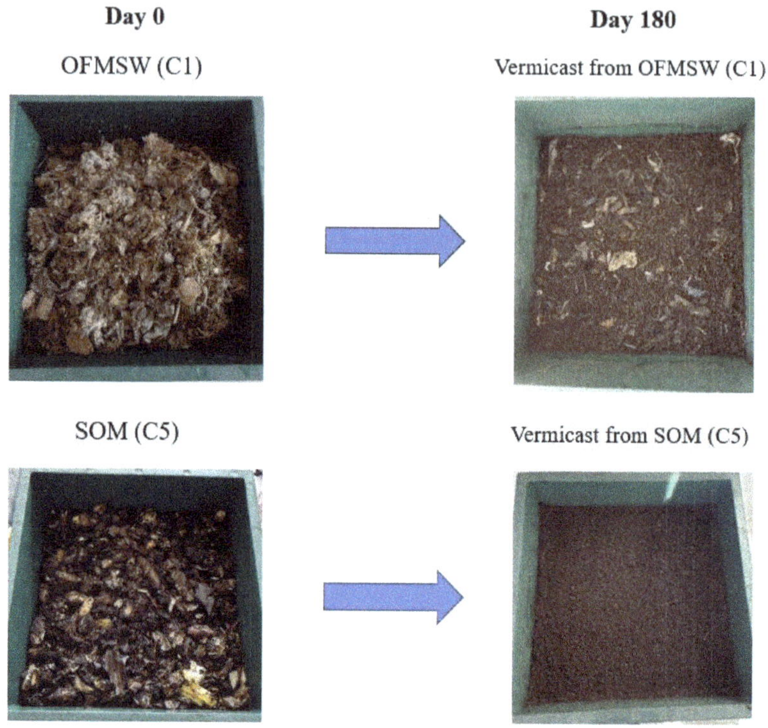

Figure 7. Transformation of the matrices after 180 days of vermicomposting.

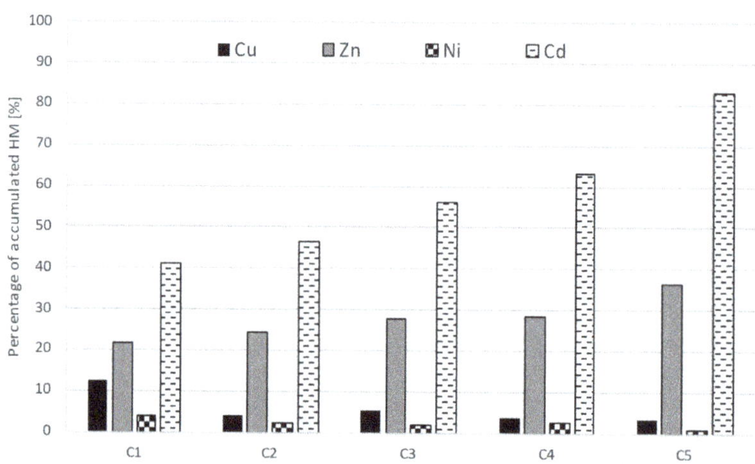

Figure 8. Percentage of accumulated heavy metals in earthworm tissues in the various tests.

Overall, the tests confirmed what was observed by [34], wherein earthworms contribute to reducing heavy metal concentrations through the bioaccumulation of varying quantities in their tissues, especially for Cu and Zn.

3.5. Future Aspects: Phyto-Growth Tests in Controlled Environment

The total concentrations of heavy metals are important for assessing potential environmental risks, especially the fraction still available in vermicast and potentially extractable, which represents a potential phytotoxic aspect for plants [43].

As previously observed, compared to the initial compost and pre-composted material, the concentrations of extractable metals in the final product have decreased significantly. In general, the proportions of available fractions to the total are almost coincident with the trends in the concentrations of metals contained in the collected liquid phase. The residual fraction can be considered "unavailable" as it is concentrated in the earthworm's tissue as a result of its own activity. Other studies have also found different metal availabilities, which, however, always decrease during the vermicomposting and maturation period [15,26,35,44].

The decrease in metal content in vermicompost occurs for at least two reasons. Firstly, vermicompost processed by earthworms generally has higher levels of humic acid fractions, which promote a stronger adsorption effect on metal–humus formation, leading to the creation of stable complexes, especially for Cu and Zn. Secondly, in line with the bioaccumulation in the tissues of earthworms, the earthworm's epithelial layer and body fluid may absorb metals during their passage through organic waste [23,44].

In conclusion, in order to assess the potential risk of transferring residual heavy metals from vermicast to plants, a series of "germination" tests should be designed to study this specific aspect. This topic is not included among the objectives of this paper; however, an experimental setup has already been prepared that will aim to test the vermicast obtained from vermicomposters C1-C5 in the future based on the observation of growth rates and the bioaccumulation of specific metals on the roots, leaves, or fruit of different species. The species analyzed will be differentiated by their ability to accumulate metals at the root level (e.g., castor bean) or on the fruit/leaves (e.g., tobacco).

4. Discussion

The study confirmed that it is possible to obtain vermicompost from different organic fractions of municipal waste, whether raw or pre-treated. In general, achievable vermicompost could potentially have better nutritive properties compared to compost and initial blends, thanks to enhanced mineralization and humification. The initial adaptation phase of worms to relatively harsh growth conditions, which can result in their mortality, is certainly crucial in this context.

On the other hand, the transformed matrices must be appropriately characterized on a case-by-case basis for their metal content. While the highest concentrations of heavy metals were still below the threshold values permitted for compost, this is not "assumed" for the leachate/worm tea obtained.

In particular, the availability of heavy metals in vermicompost is lower compared to the starting matrices and the compost obtained from a facility. The worms facilitate their transformation into an unavailable fraction through the bioaccumulation and stabilization associated with humus.

Therefore, vermicomposting technology could enhance nutrient quality compared to traditional compost (making the end product potentially a more effective soil amendment) and has the potential to mitigate the environmental risk of heavy metals from the disposal of solid organic waste. In any case, future research should assess the long-term impacts of field application.

Author Contributions: Conceptualization, E.L.; methodology, E.L. and G.D.B.; validation, E.L., M.G.G. and A.M.; investigation, E.L.; data curation, E.L. and M.G.G.; writing—original draft preparation, E.L.; writing—review and editing, G.D.B. and A.M.; supervision, G.D.B. and A.M. All authors have read and agreed to the published version of the manuscript.

Funding: This research received no external funding.

Data Availability Statement: No new data were created or analyzed in this study. Data sharing is not applicable to this article.

Acknowledgments: The authors would like to thank the Environmental and Sanitary Engineering Laboratory staff for their analytical contribution and the staff of ECOENNA SERVICES S.p.A. for the supply of analysis samples and analytical information on the raw samples.

Conflicts of Interest: The authors declare no conflicts of interest.

References

1. Ducasse, V.; Capowiez, Y.; Peigné, J. Vermicomposting of municipal solid waste as a possible lever for the development of sustainable agriculture. A review. *Agron. Sustain. Dev.* **2022**, *42*, 89. [CrossRef]
2. Ebrahimi, A.; Hashemi, H.; Eslami, H.; Fallahzadeh, R.A.; Khosravi, R.; Askari, R.; Ghahramani, E. Kinetics of biogas production and chemical oxygen demand removal from compost leachate in an anaerobic migrating blanket reactor. *J. Environ. Manag.* **2018**, *206*, 707–714. [CrossRef] [PubMed]
3. Eslami, H.; Ehrampoush, M.H.; Esmaeili, A.; Ebrahimi, A.A.; Salmani, M.H.; Ghaneian, M.T.; Falahzadeh, H. Efficient photocatalytic oxidation of arsenite from contaminated water by Fe_2O_3-Mn_2O_3 nanocomposite under UVA radiation and process optimization with experimental design. *Chemosphere* **2018**, *207*, 303–312. [CrossRef]
4. Garcia-Sanchez, M.; Tausnerova, H.; Hanc, A.; Tlustos, P. Stabilization of different starting materials through vermicomposting in a continuous-feeding system: Changes in chemical and biological parameters. *Waste Manag.* **2017**, *62*, 33–42. [CrossRef]
5. Yadav, K.D.; Tare, V.; Ahammed, M.M. Vermicomposting of source-separated human faeces for nutrient recycling. *Waste Manag.* **2010**, *30*, 50–56. [CrossRef]
6. Kumar, V.V.; Shanmugaprakash, M.; Aravind, J.; Namasivayam, S.K.R. Pilot scale study of efficient vermicomposting of agro-industrial wastes. *Environ. Technol.* **2012**, *33*, 975–981. [CrossRef] [PubMed]
7. Adhikary, S. Vermicompost, the story of organic gold: A review. *Agric. Sci.* **2012**, *3*, 905–917. [CrossRef]
8. Lim, S.L.; Lee, L.H.; Wu, T.Y. Sustainability of using composting and vermicomposting technologies for organic solid waste biotransformation: Recent overview, greenhouse gases emissions and economic analysis. *J. Clean. Prod.* **2016**, *111*, 262–278. [CrossRef]
9. Qian, X.; Shen, G.; Wang, Z.; Guo, C.; Liu, Y.; Lei, Z.; Zhang, Z. Co-composting of livestock manure with rice straw: Characterization and establishment of maturity evaluation system. *Waste Manag.* **2014**, *34*, 530–535. [CrossRef]
10. Rupani, P.F.; Embrandiri, A.; Ibrahim, M.H.; Shahadat, M.; Hansen, S.B.; Mansor, N.N.A. Bioremediation of palm industry wastes using vermicomposting technology: Its environmental application as green fertilizer. *3 Biotech* **2017**, *7*, 155. [CrossRef]
11. Lazcano, C.; Gomez-Brand, M.; Domínguez, J. Comparison of the effectiveness of composting and vermicomposting for the biological stabilization of cattle manure. *Chemosphere* **2008**, *72*, 1013–1019. [CrossRef] [PubMed]
12. Oluseyi, E.E.; Ewemoje, T.; Adedeji, A. Comparative analysis of pit composting and vermicomposting in a tropical environment. *Agric. Biosyst. Eng.* **2016**, *731*, 39757. [CrossRef]
13. Pigatin, L.B.F.; Atoloye, I.A.; Obikoya, O.A.; Borsato, A.V.; Rezende, M.O.O. Chemical study of vermicomposted agroindustrial wastes. *Int. J. Recycl. Org. Waste Agric.* **2016**, *5*, 55–63. [CrossRef]
14. Adi, A.; Noor, Z. Waste recycling: Utilization of coffee grounds and kitchen waste in vermicomposting. *Bioresour. Technol.* **2009**, *100*, 1027–1030. [CrossRef]
15. Bolong, N.; Saad, I. Characterization of university residential and canteen solid waste for composting and vermicomposting development. In *Green Engineering for Campus Sustainability*; Springer: Singapore, 2019; pp. 193–206. [CrossRef]
16. Feller, C.; Brown, G.G.; Blanchart, E.; Deleporte, P.; Cheryyanskii, S.S. Charles Darwin, earthworms and natural sciences: Various lessons from past to future. *Agric. Ecosyst. Environ.* **2003**, *99*, 29–49. [CrossRef]
17. Edwards, C.A.; Norman, Q.A.; Sherman, R. *Vermiculture Technology, Earthworms, Organic Waste and Environmental Management*; CRC Press: Boca Raton, FL, USA, 2011; pp. 17–19. [CrossRef]
18. Tauseef, S.M.; Abbasi, T.; Banupriya, D.; Vaishnavi, V.; Abbasi, S.A. HEVSPAR: A novel vermireactor system for treating paper waste. *Off. J. Pat. Off.* **2013**, *24*, 12726.
19. Khwairakpam, M.; Bhargava, R. Vermitechnology for sewage sludge recycling. *J. Hazard. Mater.* **2009**, *161*, 948–954. [CrossRef] [PubMed]
20. Suthar, S. Recycling of agro-industrial sludge through vermitechnology. *Ecol. Eng.* **2010**, *36*, 1028–1036. [CrossRef]
21. Hait, S.; Tare, V. Transformation and availability of nutrients and heavy metals during integrated composting–vermicomposting of sewage sludges. *Ecotoxicol. Environ. Saf.* **2012**, *79*, 214–224. [CrossRef]
22. Gupta, R.; Garg, V. Stabilization of primary sewage sludge during vermicomposting. *J. Hazard. Mater.* **2008**, *153*, 1023–1030. [CrossRef]
23. Suthar, S.; Singh, S. Feasibility of vermicomposting in biostabilization of sludge from a distillery industry. *Sci. Total Environ.* **2008**, *394*, 237–243. [CrossRef]
24. Ngo, P.T.; Rumpel, C.; Dignac, M.F.; Billou, D.; Duc, T.T.; Jouquet, P. Transformation of buffalo manure by composting or vermicomposting to rehabilitate degraded tropical soils. *Ecol. Eng.* **2011**, *37*, 269–276. [CrossRef]

25. Wong, J.; Selvam, A. Speciation of heavy metals during co-composting of sewage sludge with lime. *Chemosphere* **2006**, *63*, 980–986. [CrossRef] [PubMed]
26. Singh, J.; Kalamdhad, A.S. Concentration and speciation of heavy metals during water hyacinth composting. *Bioresour. Technol.* **2012**, *124*, 169–179. [CrossRef] [PubMed]
27. Alshehrei, F.; Ameen, F. Vermicomposting: A management tool to mitigate solid waste. *Saudi J. Biol. Sci.* **2021**, *28*, 3284–3293. [CrossRef] [PubMed]
28. Sun, F.S.; Yu, G.H.; Zhao, X.Y.; Polizzotto, M.L.; Shen, Y.J.; Zhou, H.B.; Zhang, X.; Zhang, J.C.; He, X.S. Mechanisms of potentially toxic metal removal from biogas residues via vermicomposting revealed by synchrotron radiation-based spectromicroscopies. *Waste Manag.* **2020**, *113*, 80–87. [CrossRef]
29. Sanchez-Monedero, M.A.; Serrami, N.; Civantos, C.G.O.; Fernandez-Hernandez, A.; Roig, A. Greenhouse gas emissions during composting of two-phase olive mill wastes with different agroindustrial by-products. *Chemosphere* **2010**, *81*, 18–25. [CrossRef]
30. Rashtbari, M.; Hossein Ali, A.; Ghorchiani, M. Effect of Vermicompost and Municipal Solid Waste Compost on Growth and Yield of Canola under Drought Stress Conditions. *Commun. Soil Sci. Plant Anal.* **2020**, *51*, 2215–2222. [CrossRef]
31. Sharma, K.; Garg, V.K. Vermicomposting of waste: A zero-waste approach for waste management. In *Sustainable Resource Recovery and Zero Waste Approaches*; Elsevier: Amsterdam, The Netherlands, 2019; pp. 133–164. [CrossRef]
32. Violante, P. Metodi di analisi chimica del suolo. In *Metodi Analitici per L'agricoltura*; Franco Angeli Collana: Milano, Italy, 2000; ISBN 8846422406.
33. Pramanik, P.; Chung, Y.R. Changes in fungal population of fly ash and vinasse mixture during vermicomposting by *Eudrilus eugeniae* and *Eisenia fetida*: Documentation of cellulase isozymes in vermicompost. *Waste Manag.* **2011**, *31*, 1169–1175. [CrossRef]
34. Song, X.; Liu, N.; Wu, D.; Qi, L.; Ye, C.; Jiao, J.; Hu, F. Heavy metal and nutrient changes during vermicomposting animal manure spiked with mushroom residues. *Waste Manag.* **2014**, *34*, 1977–1983. [CrossRef]
35. Yadav, A.; Garg, V. Recycling of organic wastes by employing *Eisenia fetida*. *Bioresour. Technol.* **2011**, *102*, 2874–2880. [CrossRef]
36. Arancon, N.Q.; Edwards, C.A. The use of vermicomposts as soil amendments for production of field crops. In *Vermiculture Technology: Earthworms, Organic Wastes, and Environmental Management*; Edwards, C.A., Arancon, N.Q., Sherman, R.L., Eds.; CRC Press: Boca Raton, FL, USA, 2011; pp. 129–151. [CrossRef]
37. Lukkari, T.; Teno, S.; Väisänen, A.; Haimi, J. Effects of earthworms on decomposition and metal availability in contaminated soil: Microcosm studies of populations with different exposure histories. *Soil Biol. Biochem.* **2006**, *38*, 359–370. [CrossRef]
38. Abbasi, S.A.; Nayeem-Shah, M.; Abbasi, T. Vermicomposting of phytomass: Limitations of the past approaches and the emerging directions. *J. Clean. Prod.* **2015**, *93*, 103–114. [CrossRef]
39. Furlong, C.; Rajapaksha, N.S.; Butt, K.R.; Gibson, W.T. Is composting worm availability the main barrier to large-scale adoption of worm-based organic waste processing technologies? *J. Clean. Prod.* **2017**, *164*, 1026–1033. [CrossRef]
40. Gomez-Brand, M.; Martínez-Cordeiro, H.; Domínguez, J. Changes in the nutrient dynamics and microbiological properties of grape marc in a continuous-feeding vermicomposting system. *Waste Manag.* **2021**, *135*, 1–10. [CrossRef] [PubMed]
41. Hu, X.; Zhang, T.; Tian, G.; Zhang, L.; Bian, B. Pilot-scale vermicomposting of sewage sludge mixed with mature vermicompost using earthworm reactor of frame composite structure. *Sci. Total Environ.* **2021**, *767*, 144217. [CrossRef]
42. Zheng, G.D.; Gao, D.; Chen, T.B.; Luo, W. Stabilization of nickel and chromium in sewage sludge during aerobic composting. *J. Hazard. Mater.* **2007**, *142*, 216–221. [CrossRef]
43. Walter, I.; Martinez, F.; Cala, V. Heavy metal speciation and phytotoxic effects of three representative sewage sludges for agricultural uses. *Environ. Pollut.* **2006**, *139*, 507–514. [CrossRef]
44. Singh, J.; Kalamdhad, A.S. Assessment of bioavailability and leachability of heavy metals during rotary drum composting of green waste (Water hyacinth). *Ecol. Eng.* **2013**, *52*, 59–69. [CrossRef]

environments

MDPI

Article

Spent Coffee Grounds (SCGs) as a Soil Amendment: The Effects of Composting Time on Early Sunflower Development

Catherine L. Sinclair [1], Peter J. Irga [2], Gabrielle Duani [1] and Fraser R. Torpy [1,*]

[1] Plants and Environmental Quality Research Group, School of Life Sciences, University of Technology Sydney, Ultimo, NSW 2007, Australia; catherine.sinclair77@gmail.com (C.L.S.); gabrielle.duani@uts.edu.au (G.D.)
[2] Plants and Environmental Quality Research Group, School of Civil and Environmental Engineering, University of Technology Sydney, Ultimo, NSW 2007, Australia; peter.irga@uts.edu.au
* Correspondence: fraser.torpy@uts.edu.au

Abstract: The unsustainability of current and future agricultural use has led to increased interest in using organic waste products, such as spent coffee grounds (SCGs) and compost, as alternative nutrient supplements. While SCGs are initially phytotoxic, upon composting, they can be utilised as a soil amendment; however, the minimum length of composting time required is not well known. Two glasshouse experiments were conducted to identify the concentration where raw SCGs are toxic to sunflower seedling growth and to assess the age and concentration at which composted SCGs are most effective for crop soil addition. Both raw and composted SCG substrates demonstrated higher water-holding capacities and electrical conductivity levels than commercial soil mix, though differences were observed in pH, with raw SCGs being acidic and composted SCGs being neutral. Concentrations of raw SCGs $\geq$ 35% caused large reductions in germination, plant height, cotyledon and true leaf emergence. SCGs composted for 6 months were non-toxic for sunflower seedlings and most effective at $\geq$35%, which could potentially reduce composting times by 50%. The addition of raw or composted SCGs also reduced the root/shoot ratio, though the cause and effects are currently unknown.

Keywords: spent coffee grounds; circular economy; phytotoxicity; composting; nature-based solutions

1. Introduction

Fertilisers are essential for maintaining crop yield in contemporary agriculture [1]. Of the nutrients provided by fertilisers, nitrogen is generally the main limiting soil nutrient for plants provided by these amendments [2,3].

Chemical fertilisers are unable to replace all nutrients lost from agricultural soils from repeated harvests [4], making the indefinite use of chemical fertilisers unsustainable. Long-term reductions in the quality of agricultural products have been recorded [5], and waterway eutrophication from nutrient runoff has been well described [6,7], which may have consequences for human health [6,7]. Furthermore, fertiliser prices can serve as an economic sink [8], having increased drastically due to recent geopolitical changes [6,8] and fossil fuel costs. Fertilisers also account for 2.1% of global greenhouse gas (GHG) emissions [9].

As a result of the growing unsustainability of synthetic fertiliser use [10], there has been increasing interest in the use of organic waste products. These wastes, such as garden waste, food waste and biosolids [11], can reduce environmental costs and waste management if properly utilised [12]. Composting facilitates the breakdown of these wastes into plant-usable forms and the stabilisation of organic materials [11], while also reducing potentially harmful bacteria, parasites and viruses [13]. Organic wastes provide a range of organic compounds and minerals suitable for agriculture [14], making them of growing interest as a soil fertiliser amendment [13]. Despite these benefits, disconnect between the agricultural and waste management sectors, as well as the absence of a general business model, have restricted large-scale prevalence [13]. As such, providing evidence to support the value of compost for agricultural use is hoped to help reduce these barriers.

An organic waste of current interest in research is spent coffee grounds (SCGs), produced as waste from the coffee brewing process [15,16]. Coffee is one of the most consumed beverages globally [17] and one of the most traded commodities [15], with its production serving as a major source of revenue and employment for developing countries such as Brazil, Vietnam and Columbia [18]. This large-scale production and consumption results in large amounts of SCGs, equating to 650 kg from each tonne of coffee produced [15,16], with the majority of globally produced SCGs (54.2%) currently discarded into landfill [19]. With at least six million tonnes of SCGs produced worldwide from just instant coffee [20], this raises landfill waste and costs and produces GHG [20]. Utilising spent coffee grounds as a soil amendment not only benefits plant growth through germination stimulation and increased leaf number [17,21] but also improves soil properties through improvements in nutrient availability, soil water-holding capacities and soil pH [22]. Furthermore, the use of SCGs keeps them out of the waste system and also supports the circular economy concept [13], which aims to minimise waste and repurpose resources for current and future sustainable development. The disposal of raw SCGs directly after use can cause potential soil and water contamination [17] due to the release of components such as phenols, caffeine and tannins [22]. Composting SCGs is known to break down these toxic components [23], producing a material suitable for plant growth [24]. Currently, 34.1% of global SCGs are utilised as composted soil amendments [17]. Composted SCGs increase plant biomass and production, as well as enhancing plant health [25], especially in combination with other composted waste products [18]. However, composting raises challenges: CO_2 is released during the composting process [26], and it is time-consuming, meaning that SCGs are typically unusable for 12 months after collection. Means by which the composting process could be shortened whilst still reducing phytotoxicity, or otherwise simplified, would be of significant benefit.

The effects of raw SCGs on plant growth are mostly attributed to their phytotoxic components: caffeine, tannins and chlorogenic acid [27]. Most studies have cited negative effects on plant growth [23], particularly at higher concentrations, through reductions in plant growth and biomass [22]. However, some studies have citied positive effects on plant growth, although only at concentrations under 10% raw SCG [21,28]. Despite negative effects of plant growth being observed most commonly at high raw SCG concentrations, a definitive concentration at which toxicity is present has yet to be identified.

SCGs have several properties beneficial for plant growth. Both raw and composted SCGs have high water-holding capacities and high levels of organic matter and macronutrients [28]. There are, however, notable differences in properties between raw and composted SCGs, with raw SCGs having lower pH values [22] and electrical conductivity (ranging from 2600 to 4500 µS [21]). Composted SCGs have a pH ranging from slightly acidic to neutral [25], making them suitable for many crop species; however, they also have much higher electrical conductivity than raw SCGs, which has potential to negatively influence plant growth [17]. The biggest difference in properties between raw and composted SCGs, however, is the presence of toxic components in the former. Composting has been shown to significantly reduce these components [23] to levels where they show no effects on plant growth [27], which offers potential for the growth of agricultural plants such as sunflowers (Helianthus annuus). Sunflowers are fast growing, agriculturally relevant and adaptable to a variety of different climate and soil conditions [29]. Furthermore, sunflowers are an important oilseed crop [30], and sunflower stalks are finding growing use as a bio-based building material, showing similar properties to other bio-based composites including industrial hemp [31].

Despite their wide availability, the agricultural use of SCGs has been restricted to date due to the general perception of their phytotoxicity, along with an aversion to time and costs associated with composting, which can range between 12 months and up to

2 years [32] before use. As such, this project aims to facilitate the broader use of SCGs in agriculture by quantifying the effects of raw and composted SCGs on plant growth at different concentrations over an extended period of time, identifying the features of any phytotoxicity that arises, and determining whether composting time could be reduced without introducing phytotoxic effects.

2. Materials and Methods

2.1. Spent Coffee Grounds' (SCGs') Composting Process

The composted spent coffee grounds (SCGs) used in these trials were collected in situ, from a farm that currently incorporates this material in commercial crops of garlic and industrial hemp. The composted SCGs used for this experiment were commercially collected from a range of Sydney cafes, including the Campos and Kua Coffee chains, which were transported to Hartley Vale Farm (−33.528, 150.244; Hartley Vale, NSW, Australia). These SCGs were composted alongside but separated from dehydrated food waste (DFW) in open windrows (Figure 1), where organics are placed in long, narrow piles and are mechanically recirculated on a bi-weekly basis. Samples were collected from fresh (SCGCF), 6-month composted (SCGCM) and 12-month composted (SCGCY) windrows. The 12-month sample was tested as this is the age at which composted SGM materials are used in practice. We selected the 6-month treatment to test whether residual toxicity was present at this stage to determine whether composting time could be reduced. Samples of SCGCF, SCGCM and SCGCY were collected from depths of 0–10 cm, 40–60 cm and >100 cm deep, respectively. Samples were transported and stored before analysis in sealed containers to minimise moisture loss. Sub-samples were sieved (710 μm) and refrigerated (4 °C) for soil profile analysis.

Figure 1. Process of open windrows.

2.2. Glasshouse Experiments

Experiments were conducted in a climate-controlled glasshouse located on the rooftop of Building 4 at the University of Technology Sydney, Australia (S 33°52′57.638″ E 151°12′2.256″), with the focus of comparing raw and composted SCG regarding their effects on plant growth.

For each glasshouse experiment, a substrate mix was created to fill the required pots for each treatment (1720 g for the raw treatments and 450 g for composted treatments). This comprised of a commercial soil mix that met the Australian Standard AS4454 for Composts, Soil conditioners and Mulches (Garden Basics Soil Mix, Pinegro Products, Goulburn NSW, Australia) and the SCG substrate (raw or composted). This commercial soil mix was selected in lieu of agricultural field substrates collected from the farm or commercial potting mix to represent a general-purpose soil with an acceptable level of plant nutrients, but without the high levels found in potting mix, or the site-specific characteristics that would typify substrates collected from a specific region. Each glasshouse trial tested the SCG concentration mixes of 0%, 2.5%, 5%, 10%, 20% and 35% SCGs by mass; however, an extra SCG concentration of 50% was added to the composted SCG treatments. This extra concentration was included due to the literature citing that the effectiveness of composted SCG was only observed at higher concentrations above 10% [33], as well as to determine whether there was an upper concentration limit of SCG compost's effectiveness. The proportion of each substrate used for each soil mix reflected the treatment (e.g., 35% raw used 65% commercial soil) and was weighed out using a scale into a prepared tray. Mixes were homogenised using a cement mixer for two minutes before being measured out evenly and compacted into the prepared 150 mm diameter pots for each treatment. The raw SCG glasshouse experiment used 10 pots for each treatment, while the composted SCG trials used 5 pots for each treatment, resulting in 15 pots total for each tested concentration across the three compost ages. These differing pot numbers were due to the initial raw SCG glasshouse trial testing two different pot sizes, tube pots (length: 5 × 5 × 12.5 cm) and circular pots (length: 9.5 × 9.5 × 14 cm), in case pot size resulted in any differences in results. As no significant differences were observed, the larger pot size was not used. After the first experiment, it was clear that the smaller sample size of 5 pots provided sufficient statistical power, and consequent experiments thus used the smaller number of independent replicates. For each glasshouse experiment, pot placement was randomised amongst treatments to avoid spatial confounding during plant growth.

Each glasshouse trial was conducted separately, with the raw treatments planted on the 5th of December 2023 and the composted treatments planted on the 25th of January 2024. Pot soils were moistened to field capacity, and two sunflower seeds were planted in each pot, 3 cm deep and 1 cm apart. For each glasshouse trial, plants were observed for four weeks after the initial planting date and were watered to field capacity thrice weekly. Observations of live plant measurements (detailed within the Results section) were conducted every weekday upon first observing germination (12 December 2023 for the raw treatments and 30 January 2024 for composted treatments). The observation period for each glasshouse experiment was 2nd of January 2024 for the raw treatments and 22nd of February for the composted treatments.

After completion of the observation period, sunflower plants were taken to the laboratory to record the final destructive plant measurements. This was conducted on the 3rd of January 2024 for raw treatments and the 23rd of February for the composted treatments. Here, soil from each pot was loosened and individual plants were removed. Each pot's soil was separated from the plants and roots and retained for analysis. Plants including roots were dried with damp paper towels and left to air dry. Upon drying, the root and shoot lengths of plants were measured and fresh weights recorded. Plants were dried in an air oven at 40 °C for 48 h, after which, dry weight was recorded. Root and shoot masses were obtained by weighing each plant after drying. Before drying, each plant was divided at the soil surface into roots and shoots, and each were individually weighed using a scale. After drying, this process was repeated for each plant, with the recorded weights being used to determine the root/shoot ratio of each plant.

2.3. Soil Analysis

Soil analysis consisted of moisture content percentage (MC%), pH, electrical conductivity (EC), total carbon (TC) and total nitrogen (TN). Each soil analysis tested the five different substrates used in the glasshouse experiments, commercial soil mix, raw SCGs, SCGCF, SCGCM and SCGCY. TC and TN tested all treatments used in the glasshouse experiments; however, this was not feasible for other soil analysis methods. For MC%, pH and EC, five samples of each substrate were tested. MC% was calculated as the difference between fresh weight (determined on samples taken directly from the glasshouse trials) and the dry weight of the same samples dried in an air oven at 40 °C to constant weight, multiplied by 100 for a percentage value. Substrate samples for the moisture content assay ranged from 10 to 20 g per sample, with five samples used per substrate.

pH was determined using the $CaCl_2$ suspension method, as it has known accuracy, following the Department of Sustainable Natural Resources Soil Survey Standard Test Method [34]. This method involved using a calibrated Milwaukee Sharp pH probe (Milwaukee Instruments, Rowville Victoria, Australia) to measure the pH of a 1:5 soil/0.01M $CaCl_2$ suspension, consisting of 10 g of air-dried soil and 50 mL of 0.01M $CaCl_2$, which had settled for 30 min after being mechanically shaken for 1 h at 15 rpm.

EC was also measured using the Department of Sustainable Natural Resources Soil Survey Standard Test Method [34]. This method used a calibrated Milwaukee Sharp EC meter (Milwaukee Instruments, Rowville Victoria, Australia) to measure the electrical conductivity of a 1:5 soil/water suspension, consisting of 10 g of air-dried soil and 50 mL of deionised water, which had been mechanically shaken for 1 h at 15 rpm.

For quantifying TC and TN, sample preparation involved air-drying samples of each substrate at 40 °C for 7 d, followed by sieving (710 µm) to improve sample homogeneity. Total carbon and nitrogen analyses were conducted using the LECO TruMac LECO TC/TN Macro Determinator (Model 630-300-900, LECO Corporation, Castle Hille NSW, Australia), which oxidises samples using an induction furnace, and quantification using an infrared detector. Before sample analysis, three blank samples were loaded into the machine to ensure the machine's calibration curve remained consistent. Due to the absence of an a priori expectation of C and N ratios, the correct standard needed to be determined before sample analysis. One sample from three of the tested substrates, these being control soil, SCGCF and SCGCM, were tested to determine the standard. From these data, it was determined that the 'wheat flour standard method' was suited for all future samples. For testing, three 200 mg samples of each substrate were measured into sample boats, weighed and loaded into the machine, which recorded TC and TN values.

2.4. Statistical Analysis

All data were collated in Microsoft Excel for statistical analysis. All data were tested for normality and homogeneity of variance before analysis and transformation (log(1 + data)) was conducted if necessary, though this was only required for first true leaf emergence for raw SCGs and the total carbon and total nitrogen data.

Data were analysed using PAST 4.14 [35]. Mean comparisons were made using ANOVA or PERMANOVA, as noted in the Results section. For data that did not consist of a mean, which include germination rates, cotyledon emergence and true leaf set emergence, data were statistically analysed using Chi Squared tests. For example, germination rate data sets had the frequency of germination compared to a null data set using Chi Squared tests. Statistical significance was accepted at $\alpha = 0.05$.

3. Results

3.1. Raw Spent Coffee Ground (SCG) Toxicity

Germination rates after 16 days were significantly different amongst the different concentrations of raw SCGs (ANOVA $p = 0.000$; Figure 2). The 0% concentration (i.e., control soil) and 5% concentration had the highest germination rates overall of 90% and 85%, respectively, after 16 days, with significant comparisons with all other treatments (Tukey's post hoc $p < 0.000$). The lowest germination was for 35% raw SCG, which was only 30% at the final measurement, confirming an inhibitory effect of raw SCG. Raw SCG concentrations from 2.5 to 20% performed worse than the 0% control, but still maintained a final germination percentage of ≥70%. Note that minor plant losses were recorded in the 0 and 10% treatments, whilst almost half of the plants in the 35% treatment did not survive 16 d.

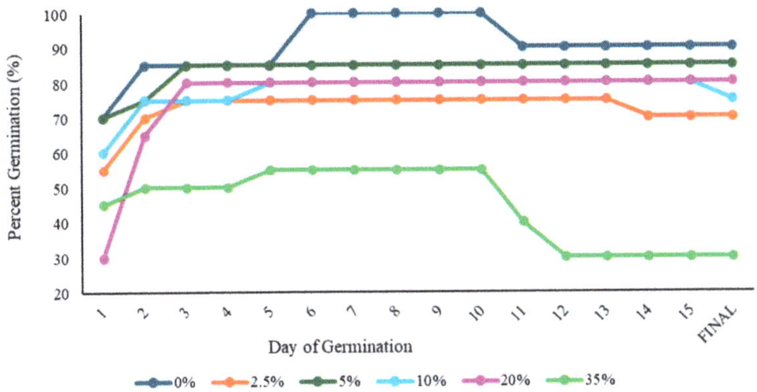

Figure 2. Differences in germination percentage amongst different raw spent coffee ground concentrations. Germination reflects the process where a plant develops from a seed into a seedling. For each day of observation, a total count of all plants that exhibited this condition was taken for each concentration and, as such, error bars were not used for this graph. Reductions in germination reflect plants that died over the course of the experiment and were thus not included in future observations.

A significant reduction in plant height was recorded in the 35% raw SCG group (Figure 3), which was almost 20% shorter than the next lowest value of 16.3 cm for the 20% treatment, with values in all other treatments ranging from 16 to 18 cm.

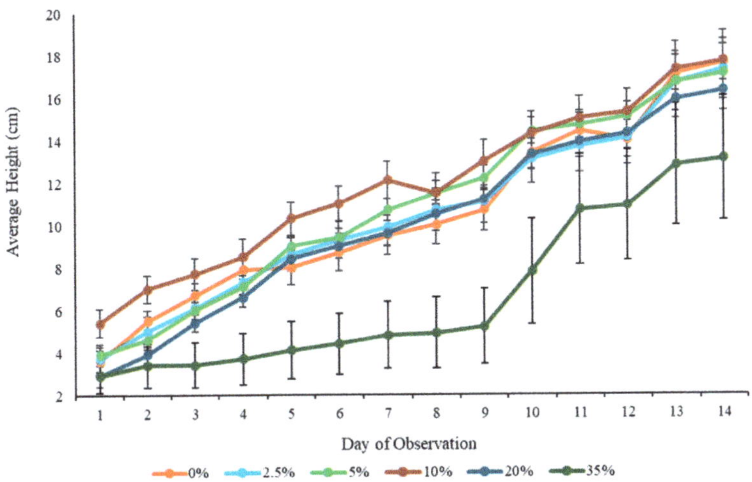

Figure 3. Differences in height amongst various raw spent coffee ground concentrations (error bars are the SEM). Height improves a plant's access to sunlight, making it able to photosynthesise easier, but can also reduce stem maintenance, making it more vulnerable to weather conditions (e.g., wind and rain).

The 35% raw SCG treatment showed the lowest cotyledon emergence of 6/20 plants (Figure 4), which was significantly different to all other concentrations ($p = 0.000$), which performed similarly.

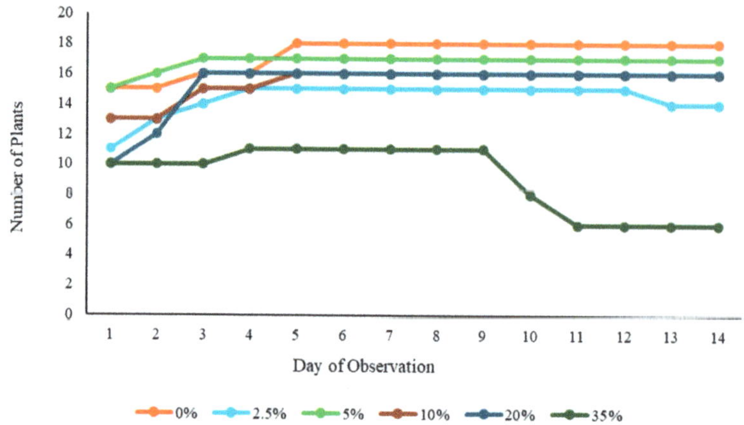

Figure 4. Cotyledon emergence of sunflowers amongst raw spent coffee ground concentrations. Cotyledon represents when plant leaves initially emerge from the seed, providing the plant with nutrients until it is able to photosynthesise. For each day of observation, a total count of all plants that exhibited this condition was taken for each concentration and, as such, error bars were not used for this graph. Reductions observed in the graph reflect plants that died over the course of the experiment and were not included in future observations.

First true leaf emergence was also significantly ($p = 0.000$) retarded in the 35% raw SCG treatment (Figure 5), with an anomalous low level also recorded in the 2.5% SCG concentration.

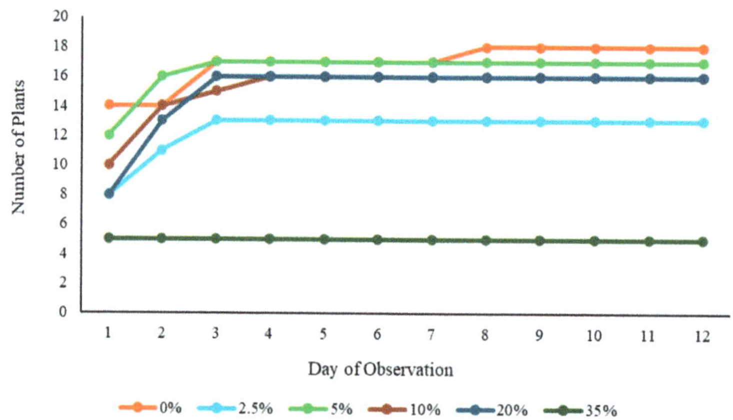

Figure 5. Number of sunflowers that experienced first true leaf emergence in soils that contained differing concentrations of raw spent coffee grounds. First true leaves are counted as the second observed leaf set on plants (as the first leaves are the cotyledon leaf set) and are the first adult leaves on a plant, allowing the plant to become a photosynthetic organism. For each day of observation, a total count of all plants that exhibited this condition was taken for each concentration and, as such, error bars were not used for this graph.

Patterns were more muted in the second true leaf emergence data set (Figure 6), with significant differences only detected between the 10% and 20% groups ($p = 0.023$) and the 10% and 35% treatments ($p = 0.009$).

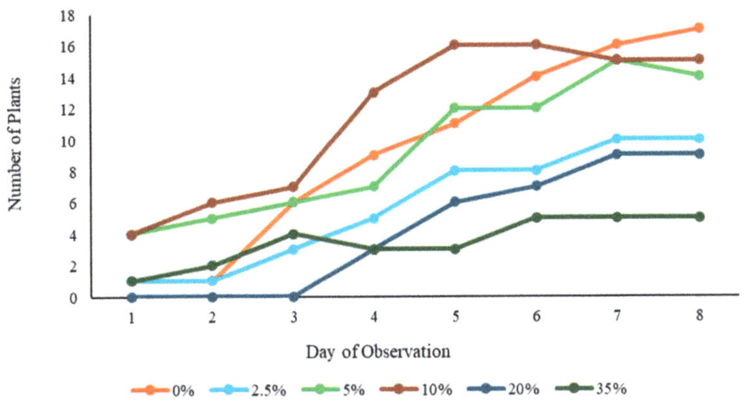

Figure 6. Second true leaf set emergence of sunflowers in soils that contained differing concentrations of raw spent coffee grounds. Second true leaves are counted as the third observed leaf set on plants (as the first leaves are the cotyledon leaf set and second are the first true leaves) and further improve a plant's ability to photosynthesise. For each day of observation, a total count of all plants that exhibited this condition was taken for each concentration and, as such, error bars were not used for this graph. Reductions observed in the graph reflect plants that died over the course of the experiment and were not included in future observations.

Plant damage was assessed subjectively and included any physical symptoms such as incompletely opened leaves, necrotic tissue, soft rots, etc. Our a priori expectation that higher concentrations of raw SCG would lead to higher numbers of damaged plants due to their higher toxicity did not occur, with no clear pattern between raw SCG treatments (ANOVA $p = 0.795$).

The total root and shoot lengths of the sunflowers were anticipated to be lower at higher concentrations due to toxicity limiting plant growth. However, no statistically significant trends were recorded for average root nor shoot length (ANOVA $p = 0.460$ and 0.259, respectively).

At the end of the experimental period, differences in the average fresh weight of sunflower seedlings amongst the different raw SCG concentrations were detected (ANOVA $p = 0.042$; Figure 7). However, Tukey's test revealed that there was only one significant comparison, between the 0% and 10% groups. Notably, the fresh weight of the 0% concentration plants was considerably lower than the 10% SCG treatment plants, potentially indicating some level of growth enhancement from raw SCG.

The patterns detected in wet weight amongst treatments were somewhat replicated in the dry weight analysis, with the highest average dry weight recorded for the 10% concentration, which was significantly higher than the 5%, 20% and 35% treatments (Tukey's $p = 0.043, 0.011, 0.026$). The 0% treatment, however, demonstrated dry weights that were not significantly lower than the 10% group ($p > 0.05$).

Root mass amongst the raw SCG treatments was highest in the 0% control, at 0.054 g, with the second highest value being 0.039 g in the 10% raw SCG plants (Figure 8). The lowest value was found in the 35% group, with an average root mass of 0.014 g. ANOVA confirmed that these findings were significant (ANOVA $p = 0.000$), with Tukey's post hoc analysis showing that all concentrations, excluding 10%, led to the development of significantly lighter root systems to the control soil treatment. There was a general tendency for higher concentrations of raw SCG to lead to smaller root masses in the seedlings.

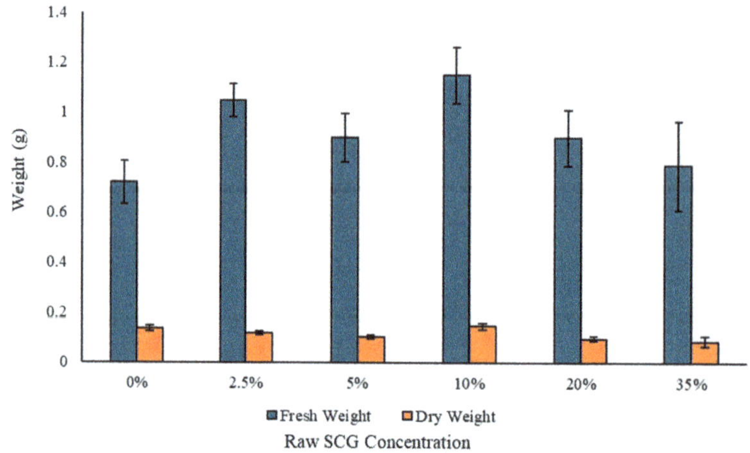

Figure 7. Comparisons of sunflower fresh weight and dry weight grown in soils containing different raw spent coffee ground concentrations (error bars are the SEM). Fresh weight is a useful measurement for evaluating the yield of plants, while dry weight is a measurement of biomass that removes any fluctuations caused by plant water content.

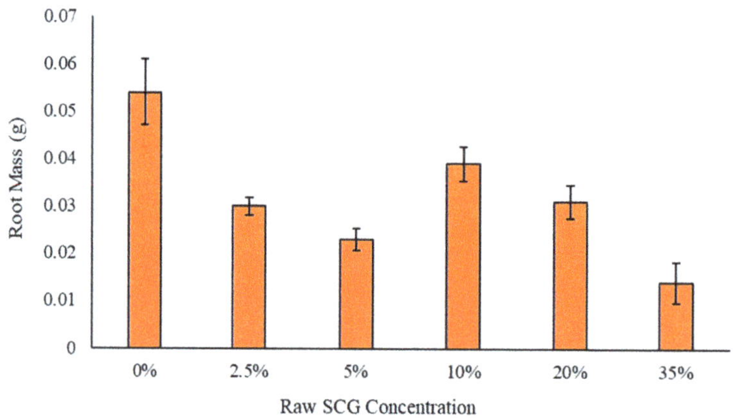

Figure 8. Root mass of sunflowers grown in different concentrations of raw spent coffee grounds (error bars are the SEM). Root mass is an important parameter to measure a plant's response to belowground environmental changes.

Shoot mass displayed a different pattern to root mass. Plants grown in 10% raw SCG had greater shoot mass than the 5%, 20% and 35% conditions (Tukey $p = 0.018$, 0.002 and 0.016), but there was no consistent pattern between SCG concentrations and the shoot mass produced (Figure 9).

Due to the unexpected patterns detected for the previous metrics, it was decided that calculating the root/shoot mass ratio may provide revealing findings (Figure 10). Any addition of raw SCGs was found to reduce the root/shoot ratio of the sunflowers (i.e., greater shoot mass at the expense of root mass), with significant differences observed between the 0% reference and all other raw SCG treatments (ANOVA, $p = 0.000$). The effect of altering root/shoot ratios in plants grown in an SCG-amended substrate has not been previously recorded and contributes to our understanding of the nature of the 'toxic' effects of raw SCG.

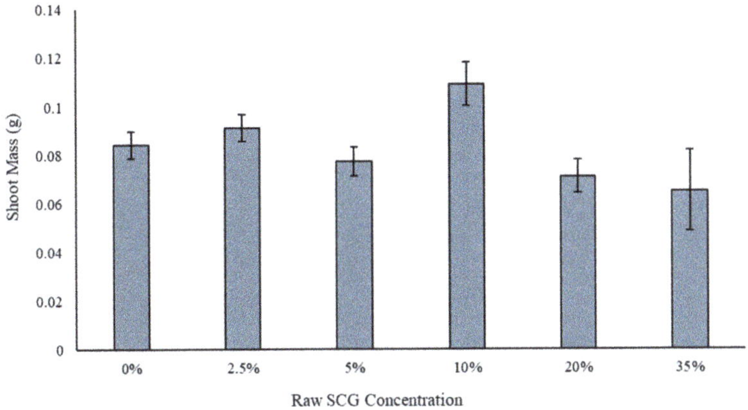

Figure 9. Shoot mass of sunflowers grown in different concentrations of raw spent coffee grounds (error bars are the SEM). Shoot mass can be used to assess a plant's response to aboveground environmental changes.

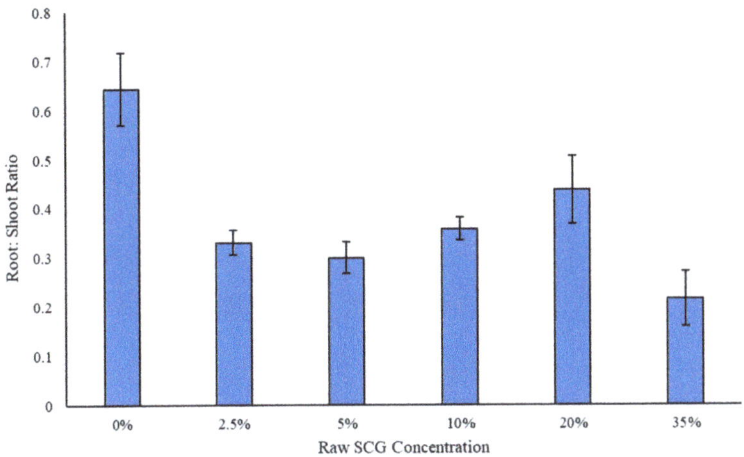

Figure 10. Comparisons of root/shoot ratio of sunflowers grown in soils that contained varying concentrations of raw spent coffee grounds (error bars are the SEM). Root/shoot ratio provides a ratio of below-ground and aboveground biomass, providing a holistic assessment of how plants respond to the surrounding environmental conditions.

Overall, live plant measurements revealed that higher concentrations of raw SCGs detrimentally affected early sunflower growth. Final values for raw SCG 0% were the highest for every live plant variable, excluding height, which was the second highest. The highest tested concentration, raw SCG 35%, displayed toxicity to sunflower growth, notably decreasing sunflower germination, plant height, cotyledon and true leaf set emergence. All other raw SCG concentrations performed worse overall than raw SCG 0% but performed significantly higher than raw 35% SCG, excluding the 2.5% and 20% concentration for second true leaf emergence. Destructive plant measurements revealed that higher concentrations, 20% and 35%, of raw SCGs resulted in reductions in dry weight, excluding 10%, where it was highest. Higher concentrations of raw SCG exhibited a growth retardation effect, resulting in smaller root masses of sunflower seedlings, also reflected in shoot mass. Interestingly, the root/shoot ratios of plants significantly decreased with any addition of raw SCG; however, the implications of this finding are currently unexplored.

3.2. Effectiveness of Spent Coffee Ground (SCG) Compost

For the first 2 days after planting, there were significant differences in the rate of germination amongst groups (χ^2 $p < 0.05$, Figure 11), with higher percentages of SCG compost associated with faster initial rates of germination (Pearson correlation between % composted SCG and germination, $p < 0.05$). However, these effects were not significant from day three onwards. These findings thus show no pathological effects of composted SCGs, but rather are evidence that higher proportions of composted SCG lead to enhanced germination rates.

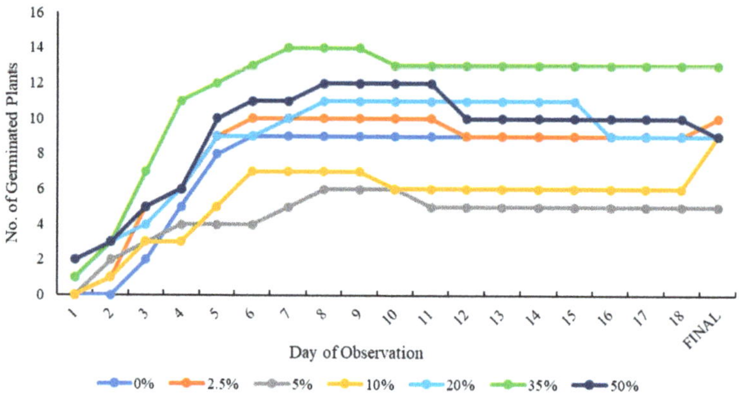

Figure 11. Number of sunflowers that germinated amongst different composted spent coffee ground concentrations. Germination reflects the process where a plant develops from a seed into a seedling. For each day of observation, a total count of all plants that exhibited this condition was taken for each concentration and, as such, error bars were not used for this graph. Reductions in germination reflect plants that died over the course of the experiment and were thus not included in future observations.

Composted SCG had no significant effects on seedling height during the 18-day growing period (Figure 12) but did affect cotyledon production.

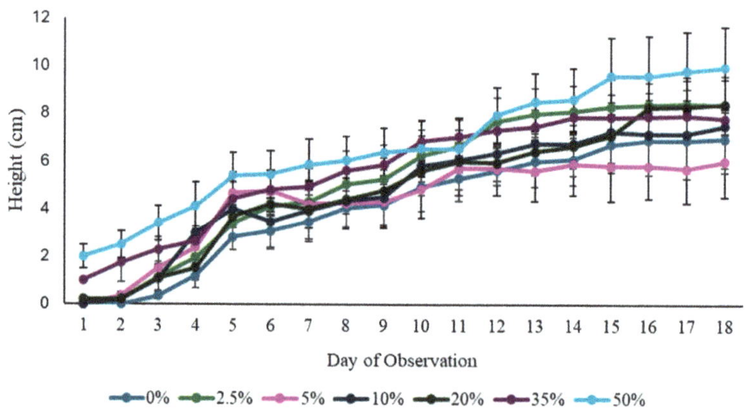

Figure 12. Plant height of sunflowers compared against different concentrations of composted spent coffee grounds (error bars are the SEM). Height improves a plant's access to sunlight, making it able to photosynthesise easier, but can also reduce stem maintenance, making it more vulnerable to weather conditions (e.g., wind and rain).

Differences in cotyledon emergence amongst composted SCG treatments were statistically significant at day 18 (ANOVA: $p = 0.000$). The highest values were achieved in the

35% concentration, while the lowest values were seen in the 5% and 10% concentrations (Figure 13). A similar pattern was recorded for true leaf production, with seedlings in the 35% treatment producing greater numbers of true leaves more rapidly than the other treatments (ANOVA $p = 0.000$; Figure 14). A similar pattern emerged for the production of the second and third true leaves, although in the case of the latter, the 50% composted SCG treatment outperformed the 35% treatment.

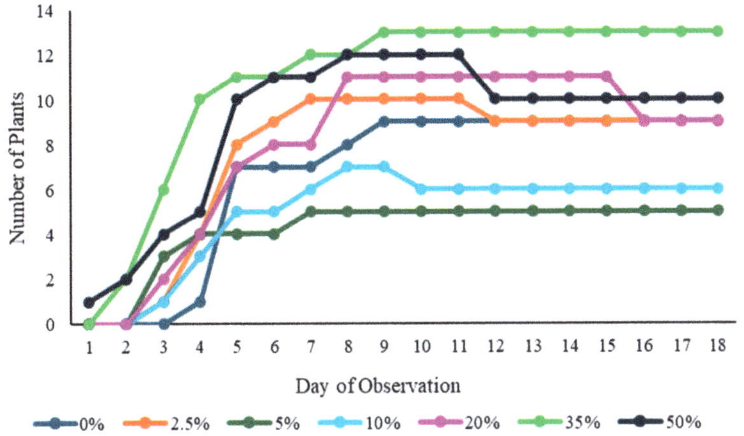

Figure 13. Cotyledon emergence of sunflowers in composted spent coffee ground soils at different concentrations. Cotyledon represents when plant leaves initially emerge from the seed, providing the plant with nutrients until it is able to photosynthesise. For each day of observation, a total count of all plants that exhibited this condition was taken for each concentration and, as such, error bars were not used for this graph. Reductions observed in the graph reflect plants that died over the course of the experiment and were not included in future observations.

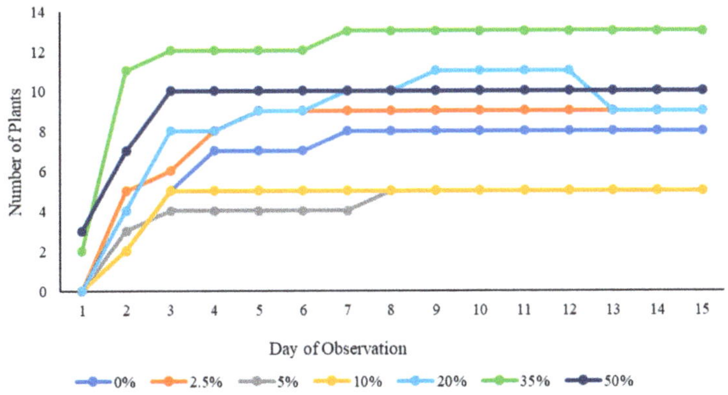

Figure 14. Sunflower first true leaf emergence in different spent coffee ground compost concentrations. First true leaves are counted as the second observed leaf set on plants (as the first leaves are the cotyledon leaf set) and are the first adult leaves on a plant, allowing the plant to become a photosynthetic organism. For each day of observation, a total count of all plants that exhibited this condition was taken for each concentration and, as such, error bars were not used for this graph. Reductions observed in the graph reflect plants that died over the course of the experiment and were not included in future observations.

Significant differences in total root length were observed amongst different SCG compost concentrations, with the 0% control treatments producing the highest average root length (Figure 15). For the other concentrations, a pattern was observed where increasing concentrations of composted SCGs resulted in shorter total root lengths in the seedlings, with this pattern statistically significant for the 10%, 35% and 50% treatments ($p = 0.016$, 0.013 and 0.035, respectively).

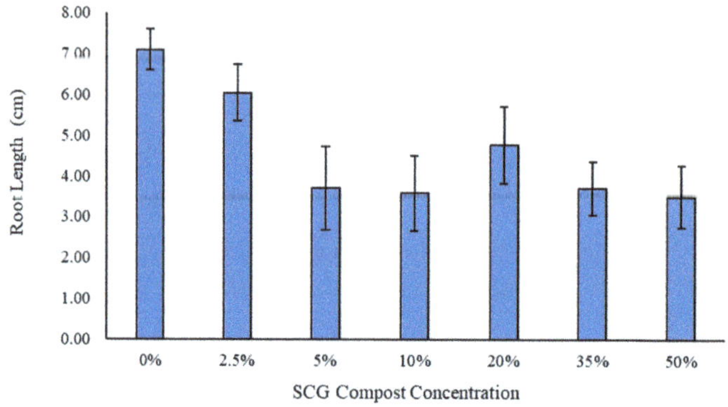

Figure 15. Differences in sunflower root length in composted spent coffee ground soils (error bars are the SEM). Root length allows plants to access and absorb higher amounts of nutrients and water from the soil.

In contrast to the findings for total root length, no significant differences were found amongst treatments for shoot length ($p = 0.244$).

There was no consistent pattern between average seedling fresh weight and the proportion of SCGC in the substrate: the only significant comparison observed was between the 50 and 35% treatments ($p = 0.020$; Figure 16). There were no significant differences in the dry weights of the seedlings amongst treatments.

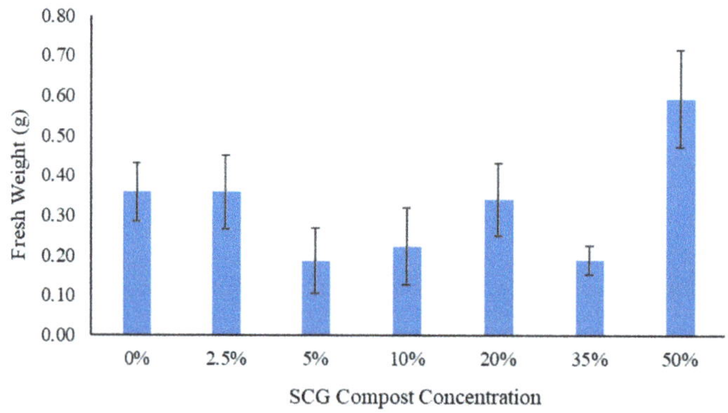

Figure 16. Comparisons of sunflower fresh weight in soils containing differing composted spent coffee ground concentrations (error bars are the SEM). Fresh weight is a useful measurement for evaluating the yield of plants.

No differences between treatments for average seedling root and shoot mass were observed, as was the case for raw SCG. There were, however, statistically significant differences ($p < 0.05$) amongst treatments for the root/shoot ratio, with the lower concentration treatments having higher root/shoot ratios (Figure 17). The relationship presented was thus similar to that shown for the raw SCG.

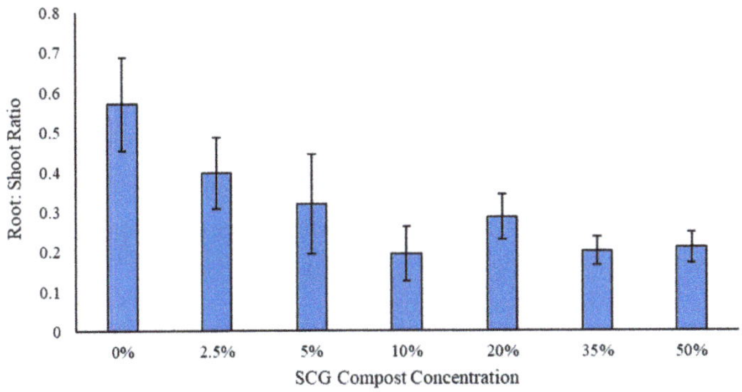

Figure 17. Sunflower root/shoot ratio in different concentrations of composted spent coffee ground soils (error bars are the SEM). The root/shoot ratio provides a ratio of belowground and aboveground biomass, providing a holistic assessment of how plants respond to the surrounding environmental conditions.

Composted SCGs demonstrated no pathological effects on sunflower seedlings, with seedlings benefiting the most from higher concentrations of composted SCGs, which were 50% and particularly 35%. Higher SCGC concentrations resulted in higher initial germination rates, increased cotyledon emergence and improvements in first, second and third true leaf emergence, with 35% performing the best in each all categories, excluding third true leaf emergence, where 50% was best. All other tested concentrations varied in performance amongst live plant variables. Destructive plant measurements showed that increased concentrations of composted SCGs reduced root lengths, and fresh weight results were inconsistent, aside from 50% having a much higher fresh weight than 35%. Like raw SCG findings, any addition of composted SCGs reduced the root/shoot ratio of plants, with lower concentrations resulting in a higher root/shoot ratio.

3.3. Soil Profile Measurements

Substrates that contained either raw or composted SCGs had a higher water-holding capacity than the reference soil (Figure 18), with a difference of 10% between soil (22.65%) and the 1-year-old SCG compost (33.45%). Raw SCGs had the highest water-holding capacity of 47.68% water content percentage (WC %), while composting 6 months or longer reduced SCG WC% significantly ($p = 0.000$).

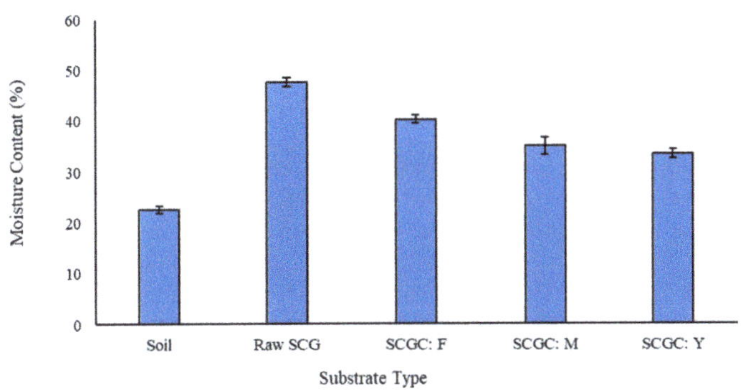

Figure 18. Moisture content percentage of differing substrates. Spent coffee ground compost (SCGC) is represented by three different decomposition states, fresh (SCGC: F), six months (SCGC: M) and >1 year (SCGC: Y) (error bars are the SEM).

Significant differences in pH amongst the substrates were observed (Figure 19), with the soil mix and raw SCGs having an acidic pH of ≈5, while composted SCGs had significantly ($p = 0.000$) higher pH values and were close to neutral. All composted treatments had statistically similar pH values ($p > 0.081$).

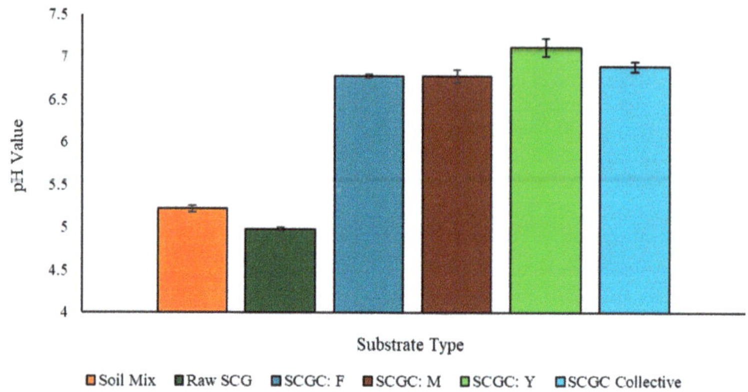

Figure 19. Differences in pH amongst different soil substrates. Spent coffee ground compost (SCGC) is rep-resented by three different decomposition states, fresh (SCGC: F), six months (SCGC: M) and >1 year (SCGC: Y) (error bars are the SEM).

The electrical conductivity of raw and composted SCGs was much higher than in the soil mix (Figure 20), with a difference of almost 500 µS between the soil mix and the SCG substrates. SCG compost at 6 months of age had the highest electrical conductivity of 1843 µS, with raw SCGs and fresh composted SCGs having similar EC levels, 1641.2 µS and 1664.4 µS, respectively. Significant differences ($p = 0.000$ in all cases) between the soil and all SCG treatments were observed.

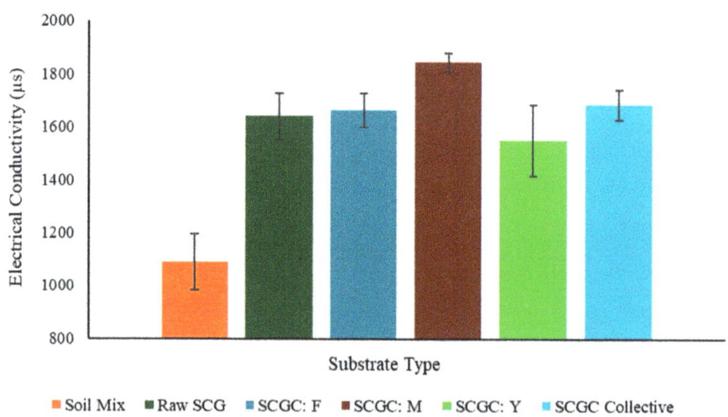

Figure 20. Comparisons of electrical conductivity between different soil substrates, measured in microsiemens (µS). Spent coffee ground compost (SCGC) is represented by three different decomposition states, fresh (SCGC: F), six months (SCGC: M) and >1 year (SCGC: Y) (error bars are the SEM).

Differences between total carbon values for the tested substrates and concentrations were significantly different ($p = 0.001$). However, post hoc testing revealed no consistent patterns related to either composting or SCG concentration and, as such, total carbon is considered unlikely to be the primary cause of the biological effects shown in the plants (Figure 21).

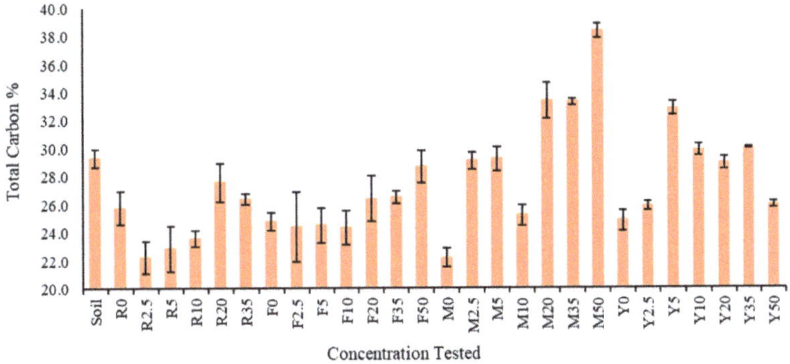

Figure 21. Total carbon measured between all substrates and concentrations. Each value is represented by a letter: raw spent coffee grounds (R) and composted spent coffee grounds at the decomposition stages of fresh (F), 6 months (M) and >1 year (Y). Each value's number represents the different tested concentrations: 0%, 2.5%, 5%, 10%, 20%, 35% and 50% (error bars are the SEM).

Total nitrogen showed significant differences amongst both substrate types and concentrations ($p = 0.001$; Figure 22). For all SCG types, comparisons between higher concentrations, in particular 35% and 50%, and lower concentrations were statistically significant ($p < 0.05$). Interestingly, the total N concentrations of the 6-month composted treatment were considerably higher than the raw and >1Y composted samples.

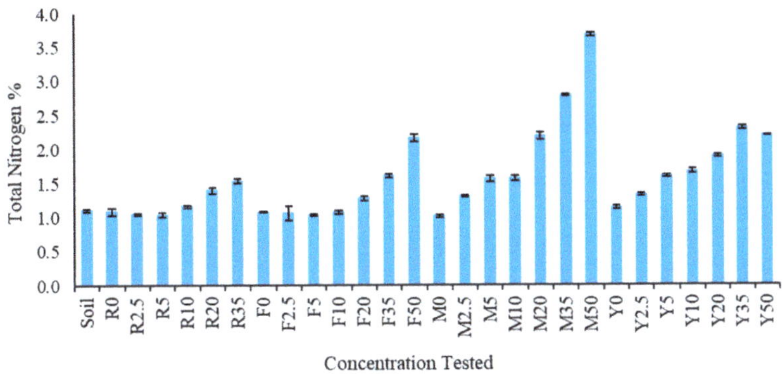

Figure 22. Total nitrogen measured all substrates and concentrations. Each value is represented by a letter: raw spent coffee grounds (R) and composted spent coffee grounds at the decomposition stages of fresh (F), 6 months (M) and >1 year (Y). Each value's number represents the different tested concentrations: 0%, 2.5%, 5%, 10%, 20%, 35% and 50% (error bars are the SEM).

Soil profiling of SCG substrates revealed that both raw and all composted SCG treatments contained a significantly higher water-holding capacity and electrical conductivity than commercial soil mix. Raw SCG pH had a similarly acidic pH to the commercial soil mix, being slightly more acidic than the soil mix. In contrast, the pH of all composted SCGs treatments was neutral. While total carbon analysis revealed no consistent patterns, total nitrogen analysis revealed that increasing concentrations within each SCG substrate increased the total nitrogen percentage of the substrate. Concentrations of 35% and 50% displayed the highest total nitrogen percentage, with the highest overall values being detected in SCGCM at concentrations of 35% and 50%.

4. Discussion

4.1. Spent Coffee Ground Substrates' Effects on Sunflower Growth

As expected a priori, raw spent coffee grounds (SCGs) were found to have a detrimental effect on plant development. All measured plant variables demonstrated that at a concentration of 35% raw SCGs, sunflowers performed significantly worse than the other tested concentrations. These findings were broadly consistent with the existing literature, showing that higher raw SCG concentrations typically result in plant growth reduction [33] with lower seed emergence [21], biomass reduction [21,23] and reduced leaf numbers [36] also being observed in the current study. Unlike other studies [28,33], which observed negative effects of raw SCGs at 10%, this study found that plants grown in the 10% and 20% concentrations performed similarly to lower concentrations, whilst in some cases outperforming them (plant height; Figure 3). Similarly, destructive plant metrics, while more variable, also indicated inferior plant performance at 35% raw SCGs. The reduced performance of plants grown in 35% raw SCGs is likely attributed to the toxic components present in raw SCGs, with both the current and previous research [37,38] finding that neither pH nor nitrogen availability were associated with negative plant health effects resulting from SCGs. The current study's scope precluded specific tests for the proposed toxic components in raw SCGs: caffeine, tannins and chlorogenic acid [39,40]. However, identifying these compounds, as well as the toxicity symptoms and concentrations caused by them, would prove beneficial towards further understanding raw SCGs' agricultural limitations. Furthermore, as sunflowers were the only plant species tested, the specific findings of this experiment may not be applicable to other plant species. As such, observing and recording live and destructive plant measurements of a variety of plant species would prove beneficial to future replications, providing a wider range of more applicable data.

Also as predicted, the current study found that higher concentrations of composted SCGs were the most effective at stimulating plant growth, with concentrations of 35% and 50% SCG compost associated with higher germination rates, improved cotyledon growth and increases in true leaf set emergence. This is consistent with previous findings that composted SCGs improved plant growth at only high concentrations [33]; however, the current findings indicate improved performance of 35% composted SCG relative to 50%, indicating that there may be a concentration limit to the useful rate at which composted SCGs can be used as a soil amendment. Composted SCGs at higher concentrations were found to reduce the root length of growing sunflowers, which indicates a growth retardation effect at higher SCG compost concentrations. This finding is novel, and it is thus recommended that further work with other crop species and longer plant growth times would be valuable for determining the scale of these effects.

A significant, novel finding of the current work was that any SCG application, either raw or composted, at any concentration affected the root/shoot ratios of the sunflower seedlings by restricting root growth without restricting aboveground phytomass. The observed effects may be attributed to the ion concentration in SCGs, as higher salinity is known to decrease root hair length and density [41]. It is not likely that the toxic components of raw SCGs are playing a role in these effects, as they were consistent between raw and composted SCGs. These effects of SCGs' effects on the root/shoot ratios of plants have not been previously documented, and so it is suggested that future research on this aspect be conducted, in particular to determine whether this is a useful or detrimental effect for agricultural purposes.

4.2. Soil Properties of Spent Coffee Ground Substrates

There were considerable differences in substrate properties amongst the substrate formulae tested within the current study. Both raw and composted SCGs had a higher water-holding capacity than the tested commercial soil product, which may be attributed to their higher levels of humic materials [42,43]. Due to water being limiting for plant growth [43], this offers potential for SCGs, particularly composted SCGs, to become a valuable soil supplement, especially for agricultural substrates that have low organic levels and low water-holding capacity, such as sandy soils. The water-holding capacity of composted SCGs decreased the longer they were decomposed, which can be attributed to the enlargement of particles during decomposition, resulting in faster soil draining [44]. There may have been some association with this effect and the reductions in root length observed in the SCG compost treatments, as particle compaction has been shown to potentially restrict root growth under some circumstances [45].

Maintaining pH within a specific band is crucial for facilitating plant macronutrient and trace metal availability [46]. Soils that are too acidic or alkaline inhibit plant growth and decrease nutrient availability, make some otherwise essential elements phytotoxic and negatively affect useful microorganisms in the soil [47]. Most plants can grow in a pH range between 5 and 8, though there is considerable variability amongst plant species. Sunflowers require soil pH ranging from slightly acidic to alkaline, with peak growth from pH = 6–6.8 [48]. Within this study, both raw SCGs and commercial soil mix had an acidic pH of ~5, making them suboptimal for sunflower growth: sunflowers grown at pH = 4.7–5.3 typically show reduced yields [49]. All the composted SCG treatments had pH = 6.7–7.1, making them suitable for sunflower growth, again supporting the value of composted SCGs as an agricultural soil amendment. Whilst the low pH of raw SCGs likely contributes to the detrimental effects of this material on plant growth, as the pH of both control soil and raw SCG substrates was similar, pH can be eliminated as the primary cause of these effects.

A specific range of substrate electrical conductivity is required for crops to take up water and nutrients. Within Australia, general agricultural irrigation limits for salinity vary amongst crops, with salt-sensitive crops handling up to 650 µS and the upper limit for moderately salt-sensitive crops being 1300 µS [50]. The electrical conductivity values seen in the tested substrates showed that SCG substrates, both raw and composted, had noticeably higher electrical conductivity (>1500 µS) compared to the soil control (1100 µS). The 600 µS difference between soil and raw SCGs is equivalent to 3.84 mg/L of dissolved ions. This would make the SCG amendments unsuitable for salt-sensitive crops, although it would be suitable for use on salt-tolerant crops such as sunflowers, which have an upper tolerance threshold of 4000–5000 µS [50].

Total nitrogen (N) levels were found to correlate to SCG concentrations, which is consistent with the existing literature [22]. SCG compost had higher N levels than raw SCGs. Six-month composted SCG at 35% and 50% had the highest N levels of 3–4%. As total N is the limiting nutrient for plant growth and is a key requirement of all fertilisers used in agriculture [51], SCGs after composting for six months would be of significant benefit for total N amendment. Comparatively, SCGs composted for >1 year had much lower total N values of around 2% at 35% and 50%, providing evidence that 6 months is a preferable time for SCG composting operations.

In contrast, total carbon (C) levels showed little variation amongst tested substrates and are unlikely to have resulted in the observed biological effects within this study. This is evidence that the properties of SCGs, whether detrimental or beneficial for plant growth, are unlikely to be due to the outright quantity of organic material provided.

4.3. Study Limitations

It was beyond the scope of the present work to assay for the postulated toxic materials in raw SCGs (e.g., caffeine and chlorogenic acid), and the restriction of the work to a single plant species constrains our ability to generalise the findings to other agricultural crops. Different plant species are known to display varying effects when exposed to SCG application and have different optimum ranges for soil parameters such as pH and electrical conductivity. Furthermore, the nutrients assessed in this study, carbon and nitrogen, are not solely responsible for plant growth, as other nutrients, such as phosphorus and potassium, are also required in varying amounts to facilitate plant growth. As such, assessing the effects of other nutrients, as well as comparing a range of plant species, will provide a broader scope for agricultural SCG application.

5. Conclusions

This study assessed the effects of raw and composted spent coffee grounds (SCGs) on plant growth, alongside assessing the differences in several critical properties of the tested substrates. Raw SCGs were found to be toxic for sunflower seedling growth at a 35% by mass concentration, with notable decreases in germination, plant height, cotyledon and true leaf set emergence. Composted SCGs assisted plant development and were most effective at higher concentrations, with 35% producing improvements in sunflower germination rate and success, as well as cotyledon and true leaf set emergence. This study also determined that SCG substrates, both raw and composted, had higher water-holding capacities and electrical conductivity levels than a reference commercial soil mix, leading to restrictions on the crops for which SCG amendments will be suitable. It was also found that the initially acidic pH of raw SCGs was ameliorated by composting.

A significant finding of this study was that the addition of SCGs was found to affect the root/shoot ratio of plants, though the causes and broader consequences of these effects are currently unknown and will require further testing. This study also found that higher concentrations of total nitrogen were found in SCGs at six months of composting, at which stage, the composted SCGs performed effectively as a soil amendment. With current agricultural practises composting SCGs for 12 months or longer, this indicates that reductions in the current SCG composting time by 50% may be possible, improving the efficiency of mass agricultural SCG use. This finding opens up the possibility of a significant contribution to circular economy: it is known that coffee consumption is approximately 12% higher in the cool months compared to in summer [52]; thus, SCG availability has a seasonal cycle. With the adoption of the reduced 6-month composting time recommended in the current work, it is possible that a matching cycle could be developed, leading to an efficient cycle of SCG–crop movement. The findings of this study indicate that SCGs can act as a valuable soil amendment after 6 months of composting time. Given the limitations of the current work, we cannot recommend its use on crops other than sunflowers with confidence without specific testing, especially for salt-sensitive species.

Author Contributions: Conceptualization, C.L.S., P.J.I. and F.R.T.; methodology, C.L.S. and F.R.T.; formal analysis, C.L.S. and F.R.T.; investigation, C.L.S. and F.R.T.; resources, C.L.S., P.J.I., G.D. and F.R.T.; data curation, C.L.S.; writing—original draft preparation, C.L.S.; writing—review and editing, C.L.S., F.R.T., G.D. and P.J.I.; visualisation, C.L.S.; supervision, F.R.T. and P.J.I.; project administration, F.R.T. and P.J.I.; funding acquisition, F.R.T. and P.J.I. All authors have read and agreed to the published version of the manuscript.

Funding: This research received no external funding.

Data Availability Statement: The raw data supporting the conclusions of this article will be made available by the authors on request.

Acknowledgments: The authors would like to acknowledge the gracious contributions of Michelle Zeibots and Hartley Vale Farm, as well as CAMPOS COFFEE PTY LTD. P.J.I. is financially supported by the ARC DECRA Scheme (DE210100755).

Conflicts of Interest: The authors declare no conflicts of interest.

References

1. Ritchie, H.; Roser, M.; Rosado, P. Fertilizers. Published Online at OurWorldInData.org. 2022. Available online: https://ourworldindata.org/fertilizers (accessed on 6 May 2024).
2. Uchida, R. Essential nutrients for plant growth: Nutrient functions and deficiency symptoms. *Plant Nutr. Manag. Hawaii's Soils* **2000**, *4*, 31–55. Available online: https://www.ctahr.hawaii.edu/oc/freepubs/pdf/pnm3.pdf (accessed on 28 April 2024).
3. Shrivastav, P.; Prasad, M.; Singh, T.B.; Yadav, A.; Goyal, D.; Ali, A.; Dantu, P.K. Role of nutrients in plant growth and development In *Contaminants in Agriculture: Sources, Impacts and Management*; Springer: Cham, Switzerland, 2020; pp. 43–59. [CrossRef]
4. Silver, W.L.; Perez, T.; Mayer, A.; Jones, A.R. The role of soil in the contribution of food and feed. *Philos. Trans. R. Soc. B* **2021**, *376*, 20200181. [CrossRef] [PubMed]

5. Jack, A. *America's Vanishing Nutrients: Decline in Fruit and Vegetable Quality Poses Serious Health and Environmental Risks*; America's Vanishing Nutrients: Becket, MA, USA, 2005; pp. 1–16. Available online: https://rockdustlocal.com/uploads/3/4/3/4/34349856/americas_vanishing_nutrients.pdf (accessed on 5 May 2024).
6. Penuelas, J.; Coello, F.; Sardans, J. A better use of fertilizers is needed for global food security and environmental sustainability. *Agric. Food Secur.* **2023**, *12*, 5. [CrossRef]
7. Tedengren, M. Eutrophication and the disrupted nitrogen cycle. *Ambio* **2021**, *50*, 733–738. [CrossRef] [PubMed]
8. Australian Government. Department of Agriculture Fisheries and Forestry. Snapshot of Global Fertiliser Trade. 2022. Available online: https://www.agriculture.gov.au/sites/default/files/documents/september-2022-snapshot-global-fertiliser-trade.pdf (accessed on 1 May 2024).
9. Menegat, S.; Ledo, A.; Tirado, R. Greenhouse gas emissions from global production and use of nitrogen synthetic fertilisers in agriculture. *Sci. Rep.* **2022**, *12*, 14490. [CrossRef]
10. Bindraban, P.S.; Dimkpa, C.; Nagarajan, L.; Roy, A.; Rabbinge, R. Revisiting fertilisers and fertilisation strategies for improved nutrient uptake by plants. *Biol. Fertil. Soils* **2015**, *51*, 897–911. [CrossRef]
11. Australian Government. Department of Climate Change, Energy, the Environment and Water. Recovering Organic Waste. 2023. Available online: https://www.dcceew.gov.au/environment/protection/waste/food-waste/recovering-organic-waste (accessed on 26 April 2024).
12. Torres-León, C.; Ramírez-Guzman, N.; Londoño-Hernandez, L.; Martinez-Medina, G.A.; Díaz-Herrera, R.; Navarro-Macias, V.; Alvarez-Pérez, O.B.; Picazo, B.; Villarreal-Vázquez, M.; Ascacio-Valdes, J.; et al. Food waste and byproducts: An opportunity to minimize malnutrition and hunger in developing countries. *Front. Sustain. Food Syst.* **2018**, *2*, 52. [CrossRef]
13. Hettiarachchi, H.; Caucci, S.; Schwärzel, K. *Organic Waste Composting Through Nexus Thinking: Practices, Policies, and Trends*; Springer Nature: Berlin/Heidelberg, Germany, 2020. [CrossRef]
14. Zhang, L.; Sun, X. Using cow dung and spent coffee grounds to enhance the two-stage co-composting of green waste. *Bioresour. Technol.* **2017**, *245*, 152–161. [CrossRef]
15. Bomfim, A.S.C.d.; de Oliveira, D.M.; Walling, E.; Babin, A.; Hersant, G.; Vaneeckhaute, C.; Dumont, M.-J.; Rodrigue, D. Spent coffee grounds characterization and reuse in composting and soil amendment. *Waste* **2023**, *1*, 2. [CrossRef]
16. Stylianou, M.; Agapiou, A.; Omirou, M.; Vyrides, I.; Ioannides, I.M.; Maratheftis, G.; Fasoula, D. Converting environmental risks to benefits by using spent coffee grounds (SCG) as a valuable resource. *Environ. Sci. Pollut. Res.* **2018**, *25*, 35776–35790. [CrossRef]
17. Hechmi, S.; Guizani, M.; Kallel, A.; Zoghlami, R.I.; Ben Zrig, E.; Louati, Z.; Jedidi, N.; Trabelsi, I. Impact of raw and pre-treated spent coffee grounds on soil properties and plant growth: A mini-review. *Clean Technol. Environ. Policy* **2023**, *25*, 2831–2843. [CrossRef]
18. Gebreeyessus, G.D. Towards the sustainable and circular bioeconomy: Insights on spent coffee grounds valorization. *Sci. Total Environ.* **2022**, *833*, 155113. [CrossRef] [PubMed]
19. Saberian, M.; Li, J.; Donnoli, A.; Bonderenko, E.; Oliva, P.; Gill, B.; Lockrey, S.; Siddique, R. Recycling of spent coffee grounds in construction materials: A review. *J. Clean. Prod.* **2021**, *289*, 125837. [CrossRef]
20. Forcina, A.; Petrillo, A.; Travaglioni, M.; di Chiara, S.; De Felice, F. A comparative life cycle assessment of different spent coffee ground reuse strategies and a sensitivity analysis for verifying the environmental convenience based on the location of sites. *J. Clean. Prod.* **2023**, *385*, 135727. [CrossRef]
21. Chrysargyris, A.; Antoniou, O.; Xylia, P.; Petropoulos, S.; Tzortzakis, N. The use of spent coffee grounds in growing media for the production of Brassica seedlings in nurseries. *Environ. Sci. Pollut. Res.* **2021**, *28*, 24279–24290. [CrossRef]
22. Hardgrove, S.J.; Livesley, S.J. Applying spent coffee grounds directly to urban agriculture soils greatly reduces plant growth. *Urban For. Urban Green.* **2016**, *18*, 1–8. [CrossRef]
23. Santos, C.; Fonseca, J.; Aires, A.; Coutinho, J.; Trindade, H. Effect of different rates of spent coffee grounds (SCG) on composting process, gaseous emissions and quality of end-product. *Waste Manag.* **2017**, *59*, 37–47. [CrossRef]
24. Kopeć, M.; Baran, A.; Mierzwa-Hersztek, M.; Gondek, K.; Chmiel, M.J. Effect of the addition of biochar and coffee grounds on the biological properties and ecotoxicity of composts. *Waste Biomass Valorization* **2018**, *9*, 1389–1398. [CrossRef]
25. Cervera-Mata, A.; Navarro-Alarcón, M.; Rufián-Henares, J.; Pastoriza, S.; Montilla-Gómez, J.; Delgado, G. Phytotoxicity and chelating capacity of spent coffee grounds: Two contrasting faces in its use as soil organic amendment. *Sci. Total Environ.* **2020**, *717*, 137247. [CrossRef]
26. Nordahl, S.L.; Preble, C.V.; Kirchstetter, T.W.; Scown, C.D. Greenhouse gas and air pollutant emissions from composting. *Environ. Sci. Technol.* **2023**, *57*, 2235–2247. [CrossRef]
27. Cervera-Mata, A.; Fernández-Arteaga, A.; Navarro-Alarcón, M.; Hinojosa, D.; Pastoriza, S.; Delgado, G.; Rufián-Henares, J. Spent coffee grounds as a source of smart biochelates to increase Fe and Zn levels in lettuces. *J. Clean. Prod.* **2021**, *328*, 129548. [CrossRef]
28. Cruz, R.; Mendes, E.; Torrinha, Á.; Morais, S.; Pereira, J.A.; Baptista, P.; Casal, S. Revalorization of spent coffee residues by a direct agronomic approach. *Food Res. Int.* **2015**, *73*, 190–196. [CrossRef]
29. Adeleke, B.S.; Babalola, O.O. Oilseed crop sunflower (*Helianthus annuus*) as a source of food: Nutritional and health benefits. *Food Sci. Nutr.* **2020**, *8*, 4666–4684. [CrossRef] [PubMed]
30. Ekins, M.G.; Hayden, H.L.; Aitken, E.A.B.; Goulter, K.C. Population structure of Sclerotinia sclerotiorum on sunflower in Australia. *Australas. Plant Pathol.* **2011**, *40*, 99–108. [CrossRef]

31. Mati-Baouche, N.; De Baynast, H.; Lebert, A.; Sun, S.; Lopez-Mingo, C.J.S.; Leclaire, P.; Michaud, P. Mechanical, thermal and acoustical characterizations of an insulating bio-based composite made from sunflower stalks particles and chitosan. *Ind. Crops Prod.* **2014**, *58*, 244–250. [CrossRef]
32. Zhao, N.; Liu, Z.; Yu, T.; Yan, F. Spent coffee grounds: Present and future of environmentally friendly applications on industries-A review. *Trends Food Sci. Technol.* **2024**, *143*, 104312. [CrossRef]
33. Gomes, T.; Pereira, J.A.; Ramalhosa, E.; Casal, S.; Baptista, P. Effect of Fresh and Composted Spent Coffee Grounds on Lettuce Growth, Photosynthetic Pigments and Mineral Composition. In VII Congreso Ibérico de Agroingeniería y Ciencias Horticolas. SECH e SEAgIng. 2014. Available online: https://core.ac.uk/download/pdf/153410245.pdf (accessed on 4 May 2024).
34. Craze, B. *Soil Survey Standard Test Method Soil Moisture Content*; Department of Sustainable Natural Resources: New South Wales, Australia, 1990.
35. Hammer, Ø.; Harper, D.A. Past: Paleontological Statistics Software Package for Educaton and Data Anlysis. *Palaeontol. Electron.* **2001**, *4*, 1. Available online: https://www.nhm.uio.no/english/research/resources/past/ (accessed on 20 April 2024).
36. Molari, M.; Dominici, L.; Comino, E. Experimenting growing media through local bio-resources valorisation: A design-oriented approach for living walls. *J. Clean. Prod.* **2024**, *436*, 140446. [CrossRef]
37. Ammar, E.E.; Rady, H.A.; Khattab, A.M.; Amer, M.H.; Mohamed, S.A.; Elodamy, N.I.; Al-Farga, A.; Aioub, A.A.A. A comprehensive overview of eco-friendly bio-fertilizers extracted from living organisms. *Environ. Sci. Pollut. Res.* **2023**, *30*, 113119–113137. [CrossRef]
38. McNutt, J. Spent coffee grounds: A review on current utilization. *J. Ind. Eng. Chem.* **2019**, *71*, 78–88. [CrossRef]
39. Horgan, F.G.; Floyd, D.; Mundaca, E.A.; Crisol-Martínez, E. Spent coffee grounds applied as a top-dressing or incorporated into the soil can improve plant growth while reducing slug herbivory. *Agriculture* **2023**, *13*, 257. [CrossRef]
40. Hollingsworth, R.G.; Armstrong, J.W.; Campbell, E. Caffeine as a repellent for slugs and snails. *Nature* **2002**, *417*, 915–916. [CrossRef] [PubMed]
41. Shelden, M.C.; Munns, R. Crop root system plasticity for improved yields in saline soils. *Front. Plant Sci.* **2023**, *14*, 1120583. [CrossRef] [PubMed]
42. Poudyal, S.; Powell, J.; Zwahlen, R.; Cardon, G. Commercially Available Products in Increase Soil Water-Holding Capacity. 2024. Available online: https://extension.usu.edu/laep/research/commercially-available-products-to-incease-soil-water-holding (accessed on 30 April 2024).
43. Bhadha, J.H.; Capasso, J.M.; Khatiwada, R.; Swanson, S.; LaBorde, C. Raising soil organic matter content to improve water holding capacity: Sl447/Ss661, 10/2017. *Edis* **2017**, *5*, 2017. [CrossRef]
44. Khan, M.O.; Klamerus-Iwan, A.; Kupka, D.; Słowik-Opoka, E. Short-term impact of different doses of spent coffee grounds, salt, and sand on soil chemical and hydrological properties in an urban soil. *Environ. Sci. Pollut. Res.* **2023**, *30*, 86218–86231. [CrossRef]
45. Mesmar, A.K.; Albedwawi, S.T.; Alsalami, A.K.; Alshemeili, A.R.; Abu-Elsaoud, A.M.; El-Tarabily, K.A.; Al Raish, S.M. The Effect of Recycled Spent Coffee Grounds Fertilizer, Vermicompost, and Chemical Fertilizers on the Growth and Soil Quality of Red Radish (Raphanus sativus) in the United Arab Emirates: A Sustainability Perspective. *Foods* **2024**, *13*, 1997. [CrossRef]
46. Cottes, J.; Saquet, A.; Palayret, L.; Husson, O.; Beghin, R.; Allen, D.; Scheiner, J.; Cabanes, C.; Guiresse, M. Effects of soil redox potential (Eh) and pH on growth of sunflower and wheat. *Arch. Agron. Soil Sci.* **2020**, *66*, 473–487. [CrossRef]
47. Rizwan, M.; Ali, S.; Rizvi, H.; Rinklebe, J.; Tsang, D.C.W.; Meers, E.; Ok, Y.S.; Ishaque, W. Phytomanagement of heavy metals in contaminated soils using sunflower: A review. *Crit. Rev. Environ. Sci. Technol.* **2016**, *46*, 1498–1528. [CrossRef]
48. Hernandez, S.T.; Carciochi, W.D.; Diovisalvi, N.V.; Izquierdo, N.; Wyngaard, N.; Barbieri, P.; Calvo, N.I.R. Does nitrogen fertilization rate, timing, and splitting affect sunflower yield and grain quality? *Crop Sci.* **2023**, *63*, 1525–1540. [CrossRef]
49. Sutradhar, A.; Lollato, R.P.; Butchee, K.; Arnall, D.B. Determining critical soil pH for sunflower production. *Int. J. Agron.* **2014**, *2014*, 894196. [CrossRef]
50. Ganf, G.; Morris, K. Salinisation and its impacts on Australia's riverine vegetation. In *Vegetation of Australian Riverine Landscapes: Biology, Ecology and Management*; CSIRO: Canberra, Australia, 2016; Volume 273. Available online: https://www.researchgate.net/publication/303520463_Salinisation_and_its_impacts_on_Australia%E2%80%99s_riverine_vegetation (accessed on 29 April 2024).
51. Körner, C. Paradigm shift in plant growth control. *Curr. Opin. Plant Biol.* **2015**, *25*, 107–114. [CrossRef]
52. Grigg, D. The worlds of tea and coffee: Patterns of consumption. *GeoJournal* **2002**, *57*, 283–294. [CrossRef]

environments

MDPI

Article

Carbon Footprint of By-Product Concentrate Feed: A Case Study

Emilio Sabia *, Ada Braghieri, Luca Vignozzi, Rosanna Paolino, Carlo Cosentino, Adriana Di Trana and Corrado Pacelli *

Department of Agricultural, Forest, Food and Environmental Sciences, University of Basilicata, 85100 Potenza, Italy; ada.braghieri@unibas.it (A.B.); luca.vignozzi@unibas.it (L.V.); rosanna.paolino@unibas.it (R.P.); carlo.cosentino@unibas.it (C.C.); adriana.ditrana@unibas.it (A.D.T.)

* Correspondence: emilio.sabia@unibas.it (E.S.); corrado.pacelli@unibas.it (C.P.)

Abstract: Using by-products in livestock feed can be an additional strategy for safeguarding land use in agriculture and reducing the environmental impact of animal production. Studies conducted on farms to assess the environmental impact of milk and meat production using life-cycle assessment (LCA) tools reveal that feeding accounts for approximately one-third. This study aimed to calculate the carbon footprint (CF) of three different concentrated feeds for livestock, both with and without the inclusion of by-products in the formulation. Three different formulations of concentrated feeds for dairy cows were developed homogeneously regarding energy content and crude protein. The LCA approach assessed CF in kg CO_2 eq.; the functional unit was 1 kg of concentrate feed. A sensitive analysis of soybean meal's association with deforestation was formulated. The concentrated feed with by-products demonstrated a lower impact on CF of 23.7% and 37.0% compared to concentrated feed with a mix of raw material and by-products, and solely with raw material, respectively. Using agricultural by-products to produce concentrated feed for livestock sectors can be an environmentally sound alternative in terms of carbon footprint.

Keywords: livestock; animal nutrition; cattle; deforestation; life-cycle assessment; by-products; global warming potential; feed industry; greenhouse gas emissions; mitigation strategy

1. Introduction

Animal-based products play a crucial role in global nutrition, serving as rich sources of both macro and micronutrients. Products from livestock contribute 18% of the world's caloric intake and 34% of the global protein consumption, while also offering vital micronutrients such as vitamin B12, iron, zinc, and calcium, which can be challenging to obtain in sufficient amounts solely from plant-based foods at the local level [1,2]. By 2050, the global population is projected to reach 9.6 billion, with 70% residing in urban areas and an average income nearly double that of today. Consequently, the demand for animal products is expected to increase, playing an essential role in ensuring global food security and nutrition [3]. The demand for animal-based products has risen significantly with the growing human population and rapid urbanisation in developing countries. By 2050, the demand for food is expected to increase by 25–70% compared to current levels [4]. Simultaneously, there is an abundance of agricultural and industrial wastes or by-products that could potentially be utilised. Agriculture currently covers approximately 38% of the world's land area [5] and is responsible for 23% of the total net anthropogenic greenhouse gas emissions [6]. A major current criticism is the low efficiency of livestock in converting feed into protein suitable for human consumption, as well as the competition between using cereals for livestock feed versus direct human consumption. However, studies estimate that 86% of livestock feed is unsuitable for human consumption. Without being utilised by livestock, a significant portion of crop residues and by-products, especially, could go to waste as the human population grows and increasingly consumes processed foods [1]. Using by-products in livestock feed can be an additional strategy for safeguarding land use in agriculture and reducing the environmental impact of animal production. For a long time, roughage and by-products from human food processing and the biofuel industry have served as feed for ruminants. In the diets of high-producing dairy cows, it is common to blend by-products with cereal and legume grains in concentrate feeds, creating a nutrient-rich ration that meets the animals' dietary needs. Including cereal grains and protein-dense legumes, such as soybean meal, in livestock diets poses a potential challenge to global food security. These commodities are highly valuable as direct sources of human nutrition and are frequently grown on prime agricultural land ideally suited for cultivating crops dedicated to feeding the human population. Allocating such resources to animal feed reduces their availability for direct human consumption. It may contribute to competition for arable land, potentially exacerbating food scarcity in regions with a limited agricultural capacity [7]. In Italy, the soybean meal used for livestock is predominantly imported from Brazil, where its production is linked to significant environmental concerns. The cultivation of soybeans in this region is often associated with high greenhouse gas emissions, primarily resulting from land-use changes such as deforestation. Additionally, this practice contributes to substantial biodiversity loss, as natural habitats are destroyed or degraded to make way for agricultural expansion [8,9]. A growing supply of literature indicates that repurposing by-products for livestock feed enhances sustainability and mitigates the environmental impacts associated with livestock production [10–13]. Incorporating crop by-products as a substitute for forages and/or concentrates in livestock diets offers several potential benefits. These include allowing for forage savings, lowering feed costs, enhancing milk production, and reducing the incidence of ruminal acidosis [14]. In intensive systems, ruminant diets are generally composed of a forage-to-concentrate ratio of 60:40; developing feed formulations that are nutritionally efficient, environmentally friendly, and involve less competition with human food resources, which could be a valuable area for research and development in the feed industry. In 2023, the industrial production of concentrated feed in Italy reached 15 million tons, with the bovine sector accounting for approximately 20% of the total [15]. Studies conducted on farms to assess the environmental impact of milk and meat production using life-cycle assessment (LCA) tools reveal that feeding accounts for approximately 30% of the carbon footprint [16–18]. Off-farm production, notably the purchase and production of concentrated feeds, contributes around 10 to 15% [19,20]. This study seeks to comprehensively evaluate the carbon footprint associated with three distinct types of concentrated livestock feeds. The analysis will consider formulations that include by-products as part of their composition and those that exclude them. By assessing the environmental impact of these feed variations, the research aims to provide valuable insights into how the incorporation of by-products might influence the overall greenhouse gas emissions associated with livestock feed production. This will contribute to understanding sustainable feeding practices and their role in reducing the environmental footprint of livestock systems.

2. Materials and Methods

Using the commercial software Supermix® v. 4.0 [21], three different formulations of concentrated feeds for dairy cows were developed. The three focused feeds were classified as concentrate feed formulated exclusively using agro-industrial by-products (by-product-concentrate feed = BCF), conventional feed formulated with raw materials and by-products (mixed concentrate feed = MCF), and a concentrate formulated exclusively using raw feed materials (concentrate raw feed = CRF); the feeds were homogeneous in terms of energy content and crude protein, as can be seen in Table 1. In detail, the main ingredients that constitute the BCF were 49.1% wheat bran, 21.6% biscuit meal, and 19.5% corn gluten meal. The MCF consisted of a mix of by-products and ingredients that can also be used for human consumption, with 31.0% maise flour, 27.8% soybean meal, 22.6% wheat bran, and 13.5% barley meal. The CRF concentrate was formulated exclusively with ingredients that compete with human food consumption, with the following composition: 43.6% soybean extraction meal, 26.3% maise meal, and 26.3% barley meal. We estimated that processing a typical ton of feed ingredients in a feed mill requires 41 kWh of electricity and 73.8 MJ/ton of natural gas with regard to energy consumption [22,23]. The carbon footprint (CF), expressed in kg of CO_2 eq./kg of functional unit (FU), of the three concentrated feeds under study was determined using the commercial software openLCA 2.0.1 and the Agribalyse v. 3.0.1 database. The FU was 1 kg of concentrate feed. The ReCiPe midpoint (H) method was used [24], and the characterisation factors employed for CF were 1, 34, and 298 CO_2 eq. for CO_2, CH_4, and N_2O, respectively. Table 2 depicts the main ingredients selected from the software's database Agribalyse, which best reflect the characteristics of the production cycles and the associated environmental impacts. The Agribalyse database is developed by INRAE, based in France, and is best suited to the characteristics of the Italian region.

Table 1. Feed and chemical composition (on a dry-matter basis); the energy required for three different concentrate feeds.

Feed Ingredients (%)	BCF	MCF	CRF
Wheat bran	49.1	22.6	-
Biscuit meal	21.6	-	-
Corn gluten meal	19.5	-	-
Sunflower meal	5.8	-	-
Soybean meal	-	27.8	43.6
Maise flour	-	31.0	26.3
Barley meal	-	13.5	26.3
Flaxseed oil	1.4	2.4	1.3
Mineral mix	2.6	2.7	2.6
Energy required			
Electricity medium voltage (Kwh/ton)	41.0	41.0	41.0
Natural gas (MJ/ton)	73.8	73.8	73.8
Nutritional value (%)			
Crude protein	22.1	22.0	22.1
Crude lipid	7.4	7.0	8.0
Starch	32.6	27.5	33.0
MFU [1]	1.09	1.09	1.09

By-product-concentrate feed (BCF); mixed concentrate feed (MCF); concentrate raw feed (CRF); [1] MFU = unit/kg dry matter; milk forage unit (MFU) = 7.11 MJ of net energy for lactation.

Table 2. Main ingredients selected from the software's database Agribalyse.

Feed Ingredients	Description	Agribalyse Database
Wheat bran	Wheat bran, animal feed, at plant/FR U	Agricultural/animal feed/feed ingredients/transformation/cereal-based/imported from SimaPro v. 9.5.
Biscuit meal	Biscuit meal, and animal feed, at the retailer's gate	Agricultural/animal feed/feed ingredients/transformation/others

Table 2. *Cont.*

Feed Ingredients	Description	Agribalyse Database
Corn gluten meal	Corn gluten meal (gluten 60)	Agricultural/animal feed/feed ingredients/transformation/cereal-based
Sunflower meal	Sunflower meal and oil, with low dehulling, at the transformation plant	Agricultural/animal feed/feed ingredients/transformation/oil seed-based
Soybean meal not associated with deforestation	Soybean meal BR, crushed in France, animal feed, at a French mill, not associated with deforestation/FR U	Agricultural/animal feed/feed ingredients/transformation/legume-based
Maise flour	Maise flour at an industrial mill	Agricultural/food/transformation
Barley meal	Barley, feed grain, conventional, stored and transported, processing/FR U	Agricultural/animal feed/feed ingredients/transformation/cereal-based
Soybean meal associated with deforestation	Soybean meal BR, crushed in France, animal feed, at a French mill, associated with deforestation/FR U	Agricultural/animal feed/feed ingredients/transformation/legume-based

2.1. Wheat Bran

In detail, wheat bran is a by-product generated during the dry-milling process of common wheat (*Triticum aestivum* L.) into flour, and it plays a significant role as an agro-industrial by-product in animal feed. It is primarily composed of the outer layers of the wheat kernel, including the cuticle, pericarp, and seed coat, along with slight traces of the starchy endosperm. Additionally, wheat bran can also result as a by-product from other wheat-processing industries that involve bran removal, such as durum wheat (*Triticum durum* Desf.) milling for pasta and semolina, as well as starch and ethanol production [25,26].

2.2. Biscuit Meal

Biscuit meal is an agro-industrial by-product generated in large quantities by biscuit manufacturing industries in various industrial regions. It is a highly palatable, energy-dense feed with wheat flour, skimmed milk powder, vegetable fat, sugar, salt, and flavouring agents. Nutritional analysis of this waste meal revealed that it contains significant amounts of protein, energy, and minerals, making it suitable for supporting animal growth and performance [11,27]. In particular, since biscuit meals are derived from leftover bread, they are regarded as residual products, and therefore, no upstream emissions are attributed to their production [26].

2.3. Soybean Meal

Soybean meal is the leading protein source in animal feed, accounting for two-thirds of global protein feed production, including all significant oilseed meals and fish meals. Its nutritional value exceeds any other plant-based protein source, making it the benchmark for comparison with other protein feeds. Soybean meal is the by-product of the extraction of soybean oil. The agricultural production process of soybeans in Brazil was selected from the Agribalyse database and is not associated with deforestation. The soybeans are then transported with 14% moisture to the French port of Brest in the Brittany region [28] by boat. Crushing takes place in France (Brest), with transport from the port to the processing plant over a distance of 20 km. Before crushing, the grains are further dried using French drying, reducing the moisture content from 13% to 11%. Economic impact allocation was applied to divide the environmental impact of soybean oil extraction from soybean meal production [26].

2.4. Corn Gluten Meal

Corn gluten meal is a by-product generated during wet milling to produce maise starch and occasionally ethanol. It is a protein-dense feed with approximately 65% crude protein (dry matter). It is utilised as a source of protein, energy, and pigments in the diets of various livestock species, including fish. The production and extraction process is located in France [28], includes maise production, drying and storing, transport from the farm to the store agency, transport from the store agency to the transformation plant, and the extraction of corn starch, with 1000 kg of maise starch from 1515 kg of grain maise, 374 kg of corn gluten feed, 40 kg of corn oil, and 75 kg of corn gluten meal (gluten 60) [26].

2.5. Sunflower Oil and Sunflower Meal

Sunflower oil and sunflower meal, with low dehulling, are in France's economic allocation, based on the Olympic average of 2008–2012 prices from La Depeche. A total of 1 kg of low-dehulled sunflower grains are used for 0.505 kg of crude oil and 0.476 kg of the meal. Prices: La Depeche, Olympic average 2008–2012, sunflower crude oil; Rotterdam: EUR 871/t, Sunflower meal 32%; Lezoux: EUR 186/t [29]. Production volume: 476 kg meal and 0.505 kg oil, including the entire process: This module describes the crushing of 1 kg of lowly dehulled sunflower grains. The transport of sunflower grains from the storing agency to the crushing plant is included in the dehulling step. The products are 0.476 kg of sunflower meal with medium protein content (32%) and 0.505 kg of sunflower crude oil, with an average French plant efficiency.

2.6. Maize Flour

Maise flour (*Zea mays* L.) is one of the most important staple grains globally, especially in regions like Africa, Latin America, and Asia, and it serves as a primary animal feed in more developed nations. Maise has many applications for food (such as grain, flour, syrup, and oil) and non-food products (including cosmetics, adhesives, paints, and varnishes). Maise starch and oil are among their key derivatives. It is a central feed grain in livestock diets, valued primarily for its energy content, and is often used as a benchmark when assessing the nutritional quality of other grains. Many by-products of maise processing—such as hominy feed, bran, germs, and oil meal from flour production, corn gluten feed and meal from starch production, and distillers' dried grains and soluble from alcohol or biofuel industries—are also used in animal feed [30]. Breeders have developed numerous maise varieties tailored to specific environmental and agronomic conditions and uses. "Dent corn" is the most widely cultivated type, predominantly used for animal feed. Other varieties, like flint corn, popcorn, sweet corn, and flour corn, are more commonly used in food production. Specific cultivars have been bred to enhance their industrial or nutritional qualities, including varieties with high lysine, high tryptophan, high oil, high amylose, or low phytate content. Brown midrib maise, for example, has a lower lignin content, which makes it more digestible for livestock. The maise production in the Agribalyse database was the average of maise from 6 case studies in the west, east, centre, and south of France. Representativeness: case studies are based on reliable expertise and statistical regional data (extrapolated from the data of the 1st quartile of the agricultural population and yields = regional mean 10%). Allocation between crops: N and P from organic fertilisers; P and K are the minerals.

2.7. Barley Meal

Barley meal (*Hordeum vulgare* L.) is one of the most important cereal crops globally. In 2009, global barley grain production reached 150 million tons, making it the fourth most-produced cereal after maise, rice, and wheat [25]. Approximately 25% of the global barley cultivation area is in developing countries. Feed barley, with 85% DM, included the following processes: the inventory consists of soil cultivation, sowing, weed control, fertilisation, pest and pathogen control, irrigation, harvest, and transport to the farm. Boundaries: from harvest to harvest. Excluded processes: soil enrichment products and post-harvest operations. Representativeness: statistical data in France, based on surveys and reliable expertise. All background data originate either from eco invent 3.4, INRA, or Agribalyse.

2.8. Sensitive Analysis of Soybean Meal Association with Deforestation

The issue of global deforestation is closely linked to soybean production in countries such as Brazil [31]. This phenomenon further exacerbates the problem of global warming due to the clearing of primary forests and the subsequent change in land use to intensive farming. To highlight this phenomenon and evaluate its effects on carbon footprint, in the formulation of conventional and mixed animal feed, a scenario has been considered where soybean meal is sourced from plots of land in Brazil associated with the deforestation of the Amazon rainforest. The following two types of feed have been formulated (Table 2): mixed concentrate feed with soybean meal from deforestation (MCFSD) and conventional raw feed with soybean meal from deforestation (CRFSD).

3. Results

3.1. Carbon Footprint

The results are shown in Figure 1. The BCF has a carbon footprint of 0.29 kg CO_2 eq. per FU. The MCF was 23.7% more impactful than BCF, with a 0.38 kg CO_2 eq. value per FU. The CRF shows a value of 0.46 kg CO_2 eq. per FU, which is 37% and 17.4% more impactful than BCF and MCF, respectively. The primary pollutants are shown in Table 3. The main pollutants for the considered feeds are CO_2 from fossils, N_2O, and CH_4 from fossils. The BCF shows that 63.4% of its carbon footprint comprises CO_2 from fossil sources, followed by N_2O (33.1%) and fossil-derived CH_4 (3.5%). The MCF shows that 69.9% of its carbon footprint impact is due to CO_2 from fossil sources, 27.2% to N_2O, and the remaining 2.9% to fossil-derived CH_4. Similarly, for CRF, 71.1% of its pollutants are attributable to CO_2 from fossil sources, 26.0% to N_2O, and 3.0% to fossil-derived CH_4.

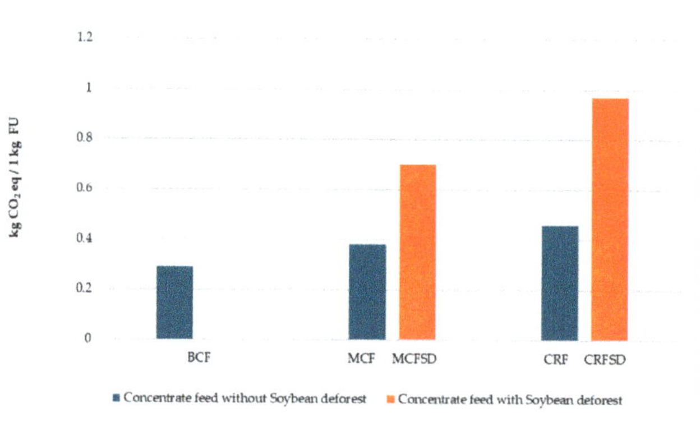

Figure 1. Carbon footprint of different concentrate feed. By-product-concentrate feed (BCF); mixed concentrate feed (MCF); conventional raw feed (CRF); mixed concentrate feed with soybean meal from deforestation (MCFSD); conventional concentrate feed with soybean meal from deforestation (CRFSD); FU = functional unit—1 kg of concentrate feed.

Table 3. Primary pollutants emitted into the air in percentages for different concentrated feeds.

Pollutants in Air (%)	BCF	MCF	MCFSD	CRF	CRFSD
CO_2 Land Transformation	0	0	42.9	0	48.5
CO_2 Fossil	62.1	68.9	40.0	69.6	33.0
N_2O	32.4	26.8	14.6	25.4	12.1
CH_4 Fossil	3.4	2.9	1.6	2.9	1.5

Cut-off, 1%; by-product-concentrate feed (BCF); mixed concentrate feed (MCF); conventional concentrate feed (CRF); mixed concentrate feed with soybean meal from deforestation (MCFSD); conventional raw feed with soybean meal from deforestation (CRFSD).

3.2. Sensitive Analysis Soybean Meal Association with Deforestation

The results of the conventional feed production scenario using soybean meal derived from deforested areas are shown in Figure 1. The BCF shows no change in its carbon footprint nor in terms of percentage of pollutants' distribution, as it does not contain soybean meal (0.29 kg CO_2 eq. per FU). The MCFSD feed has an environmental impact in terms of carbon footprint that is 58.8% higher than that of BCF and 47.7% higher than MCF, with an absolute value of 0.70 kg CO_2 eq. per FU, and the primary pollutant emitted into the air is CO_2 from land transformation, constituting 42.9%, followed by CO_2 from fossil combustion with 40.0%, and finally N_2O with 14.6%, as can be seen in Table 3. Meanwhile, the CRFSD feed demonstrates an increase in the environmental impact of 70.1% compared to BCF in terms of carbon footprint and 52.6% compared to CRF, with a value of 0.97 kg CO_2 eq. per FU, and the primary pollutant emitted into the air is CO_2 from land transformation, at 48.5%, followed by CO_2 from fossil combustion at 33.0% and finally N_2O at 12.1%, as can be seen in Table 3.

4. Discussions

Livestock production has to focus more on reducing natural resource use per unit of animal product, measured as the footprint per product, including metrics like the "water footprint", "mineral footprint", and "land footprint", whether arable or total land [32]. The carbon footprint generated by the concentrated feed based on by-products (BCF) is lower compared to a mixed concentrated feed (MCF) and a conventional concentrated feed (CRF) containing ingredients such as soybean meal and corn meal. A recent study on dairy cows demonstrated that enhancing the efficiency of on-farm feed procurement practices can significantly reduce greenhouse gas (GHG) emissions, with observed reductions ranging between 36% and 44% [33]. This finding underscores the substantial environmental benefits of improved resource management at the farm level. Similarly, research conducted in Sweden on dairy farming systems revealed that incorporating by-products into concentrated feed formulations can further contribute to environmental sustainability; in particular, the study found that substituting conventional feed components with by-product-based ingredients resulted in a 20% reduction in overall environmental impact [34]. These results suggest that optimising feed composition—particularly by leveraging agricultural by-products—presents a viable strategy for reducing the ecological footprint of dairy production. Enhancing the nutritional value of farm co-products and food waste secondary by-products generated alongside primary goods for human consumption has been proposed as a practical approach to improve resource use efficiency, minimise competition between food and feed, and promote sustainability within agricultural systems [35]. Using by-products from the farming industry fosters a circular economy that yields benefits not only from an economic standpoint but also from an environmental perspective [36]. This approach prevents these materials from ending up in landfills, where they would contribute to environmental issues if not repurposed within other industrial production cycles [37]. The current approach to soybean extraction meal is controversial, as it is technically considered a waste product, and thus a by-product, of the industrial process primarily aimed at extracting soybean oil. This oil is widely used both for human consumption and in the production of industrial and cosmetic products, as well as for energy production by biodiesel [38]. However, soybean meal, due to its high protein content, digestibility, and energy value, is one of the primary feeds used in livestock farming for meat and dairy production [39]. However, this by-product of soybean oil extraction does not compete with human food resources. The current LCA approach for estimating the carbon footprint per kilogram of soybean meal involves distributing its environmental impact through economic allocation. Since this by-product has a market, it proportionally contributes to greenhouse gas emissions. However, this method overlooks that its end-of-life destination would likely be landfilled without the livestock sector utilising this by-product, further exacerbating its environmental impact. Several studies have observed that using agricultural and food by-products does not affect the production performance of different livestock species [40–42].

5. Conclusions

Using agricultural by-products to produce concentrated feed in the livestock sector presents a potentially environmentally sustainable alternative to relying on raw vegetable materials that might otherwise compete with human food resources. This approach addresses concerns over land use and food security and offers a pathway to reduce the environmental footprint of livestock production. However, further research is essential to understand this practice's implications fully. Detailed evaluations are needed to investigate the digestibility properties of feeds formulated exclusively from agro-industrial by-products, their effects on animal performance metrics, and the quality of livestock products derived from such diets. Additionally, a thorough examination of the role of soybean meal in livestock nutrition and its associated environmental impacts is critical for developing comprehensive strategies to enhance sustainability in the sector. These studies will help inform evidence-based decisions and promote more eco-friendly feeding practices in animal agriculture.

Author Contributions: Conceptualization, E.S., A.B. and C.P.; methodology, E.S., A.B. and C.P.; software, E.S.; validation, E.S., A.B. and C.P.; formal analysis, E.S., A.B. and C.P.; investigation, E.S. and C.P.; resources, C.P.; data curation, E.S., A.B. and C.P.; writing—original draft preparation, E.S., A.B. and C.P.; writing—review and editing, E.S., A.B., L.V., R.P., C.C., A.D.T. and C.P.; visualisation, E.S., A.B. and C.P.; supervision, E.S. and C.P.; project administration, C.P.; funding acquisition, C.P. All authors have read and agreed to the published version of the manuscript.

Funding: This study was carried out within the Agritech National Research Center and received funding from the European Union Next-Generation EU (PIANO NAZIONALE DI RIPRESA E RESILIENZA (PNRR)—MISSIONE 4 COMPONENTE 2, INVESTIMENTO 1.4—D.D. 1032 17/06/2022, CN00000022). This manuscript reflects only the authors' views and opinions; neither the European Union nor the European Commission can be considered responsible for them.

Data Availability Statement: Data are available upon request.

Acknowledgments: Sincere thanks to "FUSCO—nutriamo la natura" for their technical assistance, which has been greatly appreciated.

Conflicts of Interest: The authors declare no conflicts of interest. The funders had no role in the design of the study, in the collection, analyses, or interpretation of data, in the writing of the manuscript, or in the decision to publish the results.

References

1. Mottet, A.; De Haan, C.; Falcucci, A.; Tempio, G.; Opio, C.; Gerber, P. Livestock: On Our Plates or Eating at Our Table? A New Analysis of the Feed/Food Debate. *Glob. Food Secur.* **2017**, *14*, 1–8. [CrossRef]
2. Randolph, T.F.; Schelling, E.; Grace, D.; Nicholson, C.F.; Leroy, J.L.; Cole, D.C.; Demment, M.W.; Omore, A.; Zinsstag, J.; Ruel, M. Invited Review: Role of Livestock in Human Nutrition and Health for Poverty Reduction in Developing Countries1,2,3. *J. Anim. Sci.* **2007**, *85*, 2788–2800. [CrossRef]
3. Bruinsma, J. *The Resource Outlook for 2050: By How Much Do Land. Water and Crop Yields Need to Increase by 2050?* FAO: Rome, Italy, 2009. Available online: https://www.fao.org/3/ak971e/ak971e.pdf (accessed on 17 October 2024).
4. Hunter, M.C.; Smith, R.G.; Schipanski, M.E.; Atwood, L.W.; Mortensen, D.A. Agriculture in 2050: Recalibrating Targets for Sustainable Intensification. *BioScience* **2017**, *67*, 386–391. [CrossRef]
5. Flysjö, A.; Henriksson, M.; Cederberg, C.; Ledgard, S.; Englund, J.-E. The Impact of Various Parameters on the Carbon Footprint of Milk Production in New Zealand and Sweden. *Agric. Syst.* **2011**, *104*, 459–469. [CrossRef]
6. IPCC. *Global Warming of 1.5 °C: IPCC Special Report on Impacts of Global Warming of 1.5 °C Above Pre-Industrial Levels in Context of Strengthening Response to Climate Change, Sustainable Development, and Efforts to Eradicate Poverty*, 1st ed.; Cambridge University Press: Cambridge, UK, 2022; ISBN 978-1-00-915794-0.
7. Muscat, A.; De Olde, E.M.; De Boer, I.J.M.; Ripoll-Bosch, R. The Battle for Biomass: A Systematic Review of Food-Feed-Fuel Competition. *Glob. Food Secur.* **2020**, *25*, 100330. [CrossRef]
8. Castanheira, É.G.; Freire, F. Greenhouse Gas Assessment of Soybean Production: Implications of Land Use Change and Different Cultivation Systems. *J. Clean. Prod.* **2013**, *54*, 49–60. [CrossRef]
9. Cardoso, A.S.; Berndt, A.; Leytem, A.; Alves, B.J.R.; De Carvalho, I.D.N.O.; De Barros Soares, L.H.; Urquiaga, S.; Boddey, R.M. Impact of the Intensification of Beef Production in Brazil on Greenhouse Gas Emissions and Land Use. *Agric. Syst.* **2016**, *143*, 86–96. [CrossRef]
10. Palladini, N.M.; Gislon, G.; Sandrucci, A.; Zucali, M.; Tamburini, A.; Bava, L. Assessment of Food-Feed Competition for Producing Milk in Cow Dairy Farms. *Agric. Syst.* **2024**, *218*, 103984. [CrossRef]
11. Pinotti, L.; Luciano, A.; Ottoboni, M.; Manoni, M.; Ferrari, L.; Marchis, D.; Tretola, M. Recycling Food Leftovers in Feed as Opportunity to Increase the Sustainability of Livestock Production. *J. Clean. Prod.* **2021**, *294*, 126290. [CrossRef]
12. Van Hal, O.; De Boer, I.J.M.; Muller, A.; De Vries, S.; Erb, K.-H.; Schader, C.; Gerrits, W.J.J.; Van Zanten, H.H.E. Upcycling Food Leftovers and Grass Resources through Livestock: Impact of Livestock System and Productivity. *J. Clean. Prod.* **2019**, *219*, 485–496. [CrossRef]
13. Vastolo, A.; Calabrò, S.; Cutrignelli, M.I. A Review on the Use of Agro-Industrial CO-Products in Animals' Diets. *Ital. J. Anim. Sci.* **2022**, *21*, 577–594. [CrossRef]
14. Bradford, B.J.; Mullins, C.R. Invited Review: Strategies for Promoting Productivity and Health of Dairy Cattle by Feeding Nonforage Fiber Sources. *J. Dairy Sci.* **2012**, *95*, 4735–4746. [CrossRef] [PubMed]
15. Orsetti, T. Assemblea Assalzoo, Mangimi Trainati Dagli Avicoli Nel 2023. Available online: https://informatorezootecnico.edagricole.it/economia-mercati/assemblea-assalzoo-mangimi-trainati-avicoli-2023/ (accessed on 24 September 2024).
16. Sabia, E.; Zanon, T.; Braghieri, A.; Pacelli, C.; Angerer, V.; Gauly, M. Effect of Slaughter Age on Environmental Efficiency on Beef Cattle in Marginal Area Including Soil Carbon Sequestration: A Case of Study in Italian Alpine Area. *Sci. Total Environ.* **2024**, *918*, 170798. [CrossRef]
17. Costantini, M.; Vázquez-Rowe, I.; Manzardo, A.; Bacenetti, J. Environmental Impact Assessment of Beef Cattle Production in Semi-Intensive Systems in Paraguay. *Sustain. Prod. Consum.* **2021**, *27*, 269–281. [CrossRef]
18. Batalla, I.; Knudsen, M.T.; Mogensen, L.; Hierro, Ó.D.; Pinto, M.; Hermansen, J.E. Carbon Footprint of Milk from Sheep Farming Systems in Northern Spain Including Soil Carbon Sequestration in Grasslands. *J. Clean. Prod.* **2015**, *104*, 121–129. [CrossRef]
19. Berton, M.; Bovolenta, S.; Corazzin, M.; Gallo, L.; Pinterits, S.; Ramanzin, M.; Ressi, W.; Spigarelli, C.; Zuliani, A.; Sturaro, E. Environmental Impacts of Milk Production and Processing in the Eastern Alps: A "Cradle-to-Dairy Gate" LCA Approach. *J. Clean. Prod.* **2021**, *303*, 127056. [CrossRef]
20. Angerer, V.; Sabia, E.; König Von Borstel, U.; Gauly, M. Environmental and Biodiversity Effects of Different Beef Production Systems. *J. Environ. Manag.* **2021**, *289*, 112523. [CrossRef] [PubMed]
21. Supermix Farm Computer Systems, SuperMix. Available online: https://www.fcs.it/supermix.php (accessed on 27 January 2025).
22. Garcia-Launay, F.; Van Der Werf, H.M.G.; Nguyen, T.T.H.; Le Tutour, L.; Dourmad, J.Y. Evaluation of the Environmental Implications of the Incorporation of Feed-Use Amino Acids in Pig Production Using Life Cycle Assessment. *Livest. Sci.* **2014**, *161*, 158–175. [CrossRef]
23. Ali, B.M.; Van Zanten, H.H.E.; Berentsen, P.; Bastiaansen, J.W.M.; Bikker, P.; Lansink, A.O. Environmental and Economic Impacts of Using Co-Products in the Diets of Finishing Pigs in Brazil. *J. Clean. Prod.* **2017**, *162*, 247–259. [CrossRef]

24. Huijbregts, M.A.J.; Steinmann, Z.J.N.; Elshout, P.M.F.; Stam, G.; Verones, F.; Vieira, M.; Zijp, M.; Hollander, A.; Van Zelm, R. ReCiPe2016: A Harmonised Life Cycle Impact Assessment Method at Midpoint and Endpoint Level. *Int. J. Life Cycle Assess.* **2017**, *22*, 138–147. [CrossRef]
25. Heuzé, V.; Tran, G.; Baumont, R.; Noblet, J.; Renaudeau, D.; Lessire, M.; Lebas, F. Wheat Bran. Feedipedia is a programme by INRAE, CIRAD, AFZ, and FAO. 2015. Available online: https://feedipedia.org/node/726 (accessed on 24 October 2024).
26. Wilfart, A.; Espagnol, S.; Dauguet, S.; Tailleur, A.; Gac, A.; Garcia-Launay, F. ECOALIM: A Dataset of Environmental Impacts of Feed Ingredients Used in French Animal Production. *PLoS ONE* **2016**, *11*, e0167343. [CrossRef]
27. Bikker, P.; Jansman, A.J.M. Review: Composition and Utilisation of Feed by Monogastric Animals in the Context of Circular Food Production Systems. *Animal* **2023**, *17*, 100892. [CrossRef]
28. Nguyen, T.T.H.; Bouvarel, I.; Ponchant, P.; Van Der Werf, H.M.G. Using Environmental Constraints to Formulate Low-Impact Poultry Feeds. *J. Clean. Prod.* **2012**, *28*, 215–224. [CrossRef]
29. Zmami, M.; Ben-Salha, O. Does Oil Price Drive World Food Prices? Evidence from Linear and Nonlinear ARDL Modeling. *Economies* **2019**, *7*, 12. [CrossRef]
30. Golob, P.; Farrell, G.; Orchard, J.E. (Eds.) *Crop Post-Harvest: Science and Technology*; Blackwell Science Ltd.: Oxford, UK, 2002; Volume 1, ISBN 978-0-470-75101-5.
31. Gollnow, F.; Hissa, L.D.B.V.; Rufin, P.; Lakes, T. Property-Level Direct and Indirect Deforestation for Soybean Production in the Amazon Region of Mato Grosso, Brazil. *Land Use Policy* **2018**, *78*, 377–385. [CrossRef]
32. Flachowsky, G.; Meyer, U. Challenges for Plant Breeders from the View of Animal Nutrition. *Agriculture* **2015**, *5*, 1252–1276. [CrossRef]
33. Ma, Y.; Hou, Y.; Zhang, T.; Zhu, X.; Fang, Q.; Oenema, O. Decreasing Environmental Footprints of Dairy Production Systems through Optimization of Feed Rations and Origins. *J. Clean. Prod.* **2024**, *461*, 142637. [CrossRef]
34. Lindberg, M.; Henriksson, M.; Bååth Jacobsson, S.; Berglund Lundberg, M. Byproduct-Based Concentrates in Swedish Dairy Cow Diets—Evaluation of Environmental Impact and Feed Costs. *Acta Agric. Scand. Sect.—Anim. Sci.* **2021**, *70*, 132–144. [CrossRef]
35. Sandström, V.; Chrysafi, A.; Lamminen, M.; Troell, M.; Jalava, M.; Piipponen, J.; Siebert, S.; Van Hal, O.; Virkki, V.; Kummu, M. Food System By-Products Upcycled in Livestock and Aquaculture Feeds Can Increase Global Food Supply. *Nat. Food* **2022**, *3*, 729–740. [CrossRef] [PubMed]
36. De Pascale, A.; Arbolino, R.; Szopik-Depczyńska, K.; Limosani, M.; Ioppolo, G. A Systematic Review for Measuring Circular Economy: The 61 Indicators. *J. Clean. Prod.* **2021**, *281*, 124942. [CrossRef]
37. Fussone, R.; Cannella, S.; Dominguez, R.; Framinan, J.M. On the Bullwhip Effect in Circular Supply Chains Combining By-Products and End-of-Life Returns. *Appl. Math. Model.* **2025**, *137*, 115670. [CrossRef]
38. Li, Y.; Xu, H.; Northrup, D.; Wang, M. Effects of Soybean Varieties on Life-cycle Greenhouse Gas Emissions of Biodiesel and Renewable Diesel. *Biofuels Bioprod. Biorefining* **2023**, *17*, 449–462. [CrossRef]
39. Fischer, R.A. Definitions and Determination of Crop Yield, Yield Gaps, and of Rates of Change. *Field Crops Res.* **2015**, *182*, 9–18. [CrossRef]
40. Tretola, M.; Ottoboni, M.; Luciano, A.; Rossi, L.; Baldi, A.; Pinotti, L. Former Food Products Have No Detrimental Effects on Diet Digestibility, Growth Performance and Selected Plasma Variables in Post-Weaning Piglets. *Ital. J. Anim. Sci.* **2019**, *18*, 987–996. [CrossRef]
41. Kaltenegger, A.; Humer, E.; Stauder, A.; Zebeli, Q. Feeding of Bakery By-Products in the Replacement of Grains Enhanced Milk Performance, Modulated Blood Metabolic Profile, and Lowered the Risk of Rumen Acidosis in Dairy Cows. *J. Dairy Sci.* **2020**, *103*, 10122–10135. [CrossRef]
42. Aljerf, L.; AlMasri, N.; Prince, U. Statistical Relationship between Milk Constituents Used in Breeding Programs during Lactation: French Case Study. *Madridge J. Case Rep. Stud.* **2018**, *2*, 90–93. [CrossRef]

environments MDPI

Article

The Phytoremediation of Arsenic-Contaminated Waste by *Poa labillardieri*, *Juncus pauciflorus*, and *Rytidosperma caespitosum*

Feizia Huslina [1,2,3,*], Leadin S. Khudur [1,2], Julie A. Besedin [1,2,4], Kamrun Nahar [1,5], Kalpit Shah [1,5], Aravind Surapaneni [1,6], Pacian Netherway [4] and Andrew S. Ball [1,2]

1 Australian Research Council Training Centre for the Transformation of Australia's Biosolids Resource, Royal Melbourne Institute of Technology University, Bundoora, VIC 3083, Australia; leadin.khudur@rmit.edu.au (L.S.K.); s3962676@student.rmit.edu.au (J.A.B.); s3855513@student.rmit.edu.au (K.N.); kalpit.shah@rmit.edu.au (K.S.); aravind.surapaneni@rmit.edu.au (A.S.); andy.ball@rmit.edu.au (A.S.B.)
2 School of Science, Royal Melbourne Institute of Technology University, Bundoora, VIC 3083, Australia
3 Department of Biology, Faculty of Science and Technology, State Islamic University of Ar-Raniry, Banda Aceh 23111, Indonesia
4 Environment Protection Authority Victoria, EPA Science, Centre for Applied Sciences, Ernest Jones Drive, Macleod, Melbourne, VIC 3085, Australia; pacian.netherway@epa.vic.gov.au
5 School of Engineering, Royal Melbourne Institute of Technology University, Bundoora, VIC 3083, Australia
6 South East Water, 101 Wells Street, Frankston, VIC 3199, Australia
* Correspondence: s3825923@student.rmit.edu.au

Abstract: Phytoremediation represents a potentially effective and environmentally friendly technology to remediate arsenic (As) in mine waste soils. However, soil amendments are often required to improve phytoremediation due to depleted nutrients in mine waste. This study aims to assess the effect of biosolids biochar, applied at different rates (0%, 5%, and 10%) on As phytoremediation using three plant species: *Poa labillardieri*, *Rytidosperma caespitosum*, and *Juncus pauciflorus*. This study was conducted in a replicated greenhouse pot study using soil from an abandoned mine site. Dry plant biomass, As concentration in plants and soil, and soil microbial abundance were investigated. *Juncus pauciflorus* produced eight times more root and shoot biomass than *R. caespitosum* in soils amended with 10% biochar. The highest As uptake was also observed in *J. pauciflorus* grown in soils amended with 10% biochar (7.10 mg/plant), while *R. caespitosum* had the lowest As uptake in soils without biochar (0.16 mg/plant). In soils amended with 10% biochar, the total bacterial community decreased to approximately 8.50 $\log_{10}$ copies/g, compared to the initial soil (9.05 $\log_{10}$ copies/g), while the number of gene copies of the *nifH* gene increased, suggesting the importance of nitrogen-fixing bacteria to promote plant growth. Taguchi analysis confirmed that plant species was the key factor for As phytoremediation, followed by biochar application dose. This study showed that *J. pauciflorus* and the addition of 10% biochar was the best treatment for remediating As-contaminated mine waste, offering the potential for use commercially. Moreover, the utilisation of biochar derived from biosolids as a soil amendment for enhancing phytoremediation represents good circular economy practice to manage excessive biosolids production.

Keywords: arsenic; biochar; *Juncus pauciflorus*; mine soils; *nifH* gene; phytoremediation; Taguchi methodology

1. Introduction

In recent years, soil pollution by toxic elements has been an increasing environmental issue worldwide [1,2]. Arsenic (As) is one element that poses negative impacts to humans and the environment. Arsenic has been placed on the Substance Priority List (2022) by the Agency for Toxic Substances and Disease Registry (ATSDR) [3]. Exposure of humans to As can result in acute to chronic poisoning; exposure occurs through contaminated groundwater that is utilised as drinking water. Over 100 million people in Bangladesh, India, China, and Taiwan are impacted by As [4–7]. Some of the reported diseases include skin lesions, gastroenteritis, vascular diseases, skin cancer, and other cancers of the liver and kidneys [8].

Arsenic pollution can originate from either natural or anthropogenic sources [9]. Mining practices are one of the major anthropogenic sources; As contamination in mine soil has become a major concern worldwide [10]. Increased As concentrations (10,000 mg/kg) have been reported in soils and tailings at gold mine sites in the Amazon, a concentration that exceeds the standard value of 5.9 mg/kg, regulated by the Brazilian National Council of the Environment [11]. Yang, Xie [12] also reported that soil from a realgar mine site in China had elevated As concentrations (83–2224 mg/kg), up to 55 times higher than the Grade III values (40 mg/kg) of the China Environmental Quality Standard for Soils. In Australia, abandoned metal mine sites represent an important source of As contamination in the environment; Smith, Smith [13] stated that increased levels of As were commonly found in acid mine drainage (AMD) waste. The majority of As detected in abandoned mine soils is in the residual fraction, which contains particular types of grains of As minerals originating from mine waste that merged with the soil mineral portion [14]. Eisler [15] concluded that soil and rock waste, ore roasting, and tailings are mining residues that become sources of As pollution in mine sites; however, tailings are the greatest source of As. In mine tailings dumps, the fragmentation of sulphide minerals such as arsenopyrite (FeAsS) has been reported to produce raised concentrations of As [16,17].

Due to the severe impacts on human health and the ecosystem, it is essential to apply an effective technique to restore As-contaminated mine soils. Several methods (physical, chemical, and biological) have been applied for As-contaminated mine soils. Physical methods, such as electrokinetic remediation (EK) and turnover are beneficial in reducing As; nevertheless, EK is limited to extracting small amounts of As in mobile forms, while turnover requires expert labour and high cost and maintenance [18]. Soil immobilisation and soil washing, which are categorised as chemical methods, can be applied in the field; however, long-term monitoring and secondary pollution during soil transportation are key limitations. Biological methods (bioremediation and phytoremediation) are more environmentally friendly and cost-effective; nevertheless, bioremediation studies are frequently limited to controlled laboratory settings, despite the fact that this approach shows significant potential for substantial As removal, particularly when integrated with other treatment approaches. Phytoremediation is a technique that utilises plants to remediate contaminated soil. This approach represents an environmentally friendly way to restore As mine soils [19,20]. Phytoextraction and phytostabilisation are the two main mechanisms involved in As phytoremediation. Phytoextraction refers to the ability of plant roots to uptake contaminants from soil and accumulate them within plant biomass. Phytostabilisation is the mechanism used by plants to reduce the mobility and bioavailability of contaminants through absorption and precipitation by roots, thereby reducing metal concentrations in the rhizosphere [21,22]. Phytoremediation represents a long-term strategy, as plant growth, especially extensive root formation, is required for effective treatment [23]. However, phytoremediation has been successfully applied in As-contaminated soils, both for laboratory and field experiments [24–26]. Nevertheless, because of the limited soil organic matter and fertility associated with mine waste soils, soil amendment is paramount during phytoremediation to enhance plant growth and metal uptake [27,28]. For example, compost was utilised during As phytoremediation by *Zea mays* L. in both sandy loam- and clay loam-contaminated soils [29]. The study showed that the addition of compost (at 1% and 2.5% *v/v*) enhanced the phytoremediation of As in sandy loam soils and clay loam soils, respectively. Yang, Yang [30] applied ammonium dihydrogen phosphate ($NH_4H_2PO_4$) and citric acid to soils during As phytoremediation by *Pteris vittata* L. Both soil additives promoted plant biomass and As uptake by the plant; however, $NH_4H_2PO_4$ promoted higher As uptake in the root. Iron is another alternate amendment used by Souri, Karimi [31]. They showed that As uptake by *Isatis cappadocica* increased by 56% following the addition of iron oxide nanoparticles (nFe_3O_4 and nFe_3O_4@GSH).

Among soil amendments, biochar has been extensively utilised to improve phytoremediation and represents a promising approach to enhancing plant growth [32,33]. Biochar is a cost-effective material with a high carbon content that enhances soil properties, including pH, microbial activities, and nutrients [27]. Biochar is usually produced by pyrolysis from various sources of feedstock, such as manure and agricultural waste [34]. Biochar's role in remediation is to act as a biostimulator or a sorbent. Biochar acts as a biostimulator by changing soil properties, offering nutrients and protection for microbes, enhancing bioremediation. It also functions as a sorbent by decreasing soil contamination through the absorption of contaminants [35–37]. Previous studies used the addition of biochar to assist As uptake through phytoremediation in mine soil; for example, biochar produced from rice straw was applied to soils during the phytoremediation of As in mine soil using *Pteris vittata*. The results revealed that a low dose of biochar (1.5%) was most effective in terms of increased As accumulation in plants [38]. However, biochar added at 10% was shown to enhance biomass, As uptake, and the microbial community, resulting in a reduction in the As content in the soil [39]. Guo, Chen [33] applied different rates of biochar (2%, 5%, and 10%) to remediate As-contaminated soil from an abandoned mine site in China. The result shows that 10% biochar increased the phytoremediation efficiency of *Pteris vittata* by 11.63% compared to the control. It also showed that biochar increased the soil pH and organic matter. Moreover, the soil bacterial community changed, indicated by the increase in *Streptomyces* from 26.6 to 54.2% and *Sphingomonas* from 12.3 to 30.8%.

In any study aimed at examining the phytoremediation of contaminated soils, it is important to investigate the impact of biochar on the soil microbial community. Soil microbes play a significant role in delivering soil ecosystem services such as the biogeochemical cycling of nutrients and therefore represent an important indicator of soil health. The soil microbial community can improve plant growth by increasing nutrient uptake and protecting plants from pathogens [40,41]. In this study, the 16S rRNA gene was chosen to determine total bacterial numbers. The *nifH* gene was also assessed due to the common limitation of nitrogen in mine tailings, which often inhibits plant growth [42,43]. The abundance of *nifH* gene copies has been observed to increase as plant succession advances, such as in copper mine tailings [44].

This study aims to investigate the effect of biosolids biochar applied at different rates on As phytoremediation using three plant species: *Poa labillardieri*, *Rytidosperma caespitosum*, and *Juncus pauciflorus*. The plants were selected based on previous studies. *Poa labillardieri* and *Rytidosperma caespitosum* have been reported as metal accumulators [45,46], while *Juncus* is the plant genus growing in the study site and reported as a metal accumulator at other mine sites [47,48]. Other *Juncus* species, such as *Juncus acutus*, have been studied in terms of phytoremediation and found to be capable of As accumulation [49]. However, to date, there have been no studies investigating the utilisation of *Juncus pauciflorus*, along with *Poa labillardieri* and *Rytidosperma caespitosum*, for the phytoremediation of As from mine waste soil. Although research on the application of biochar to phytoremediation has been conducted, the combination of biosolids biochar and specific plant species in As-contaminated soils has not been studied. The use of biosolids, which are rich in nutrients like nitrogen, phosphorus, and micronutrients for producing biochar, which is then applied in remediation projects, presents an opportunity to utilise resources that may otherwise not be utilised [50]. Although biosolids are currently widely used in agriculture and land rehabilitation, emerging contaminants detected in biosolids represent a significant threat to its continued application [50]. In Australia, the production of biosolids has increased from 349,000 tonnes in 2021 to 372,000 tonnes in 2023. The conversion of biosolids to biochar represents a promising approach to manage excessive biosolids production while removing contaminants such as plastics and per- and polyfluoroalkyl substances (PFAS) [50,51].

2. Materials and Methods

2.1. Soil Sampling

Clay mine waste soils were sampled from an abandoned mine site in Bendigo, Victoria, Australia (Figure 1: coordinates: 36.76121° S, 144.24190° E) using a purposive sampling method. The sampling location was selected based on literature research, as well as the knowledge and experience of the authors. Confirmation was obtained from a preliminary site visit. The location was previously utilised by the United Pyrites Gold Extraction Co. for the treatment of mine tailings containing gold. Deebles United Pyrite Works and Edwards and Deebles were operational from around 1872 until sometime between 1907 and 1912. The area is currently unoccupied and utilised for open areas, leisure, and preservation activities. Arsenic has been reported as one of the potential contaminants at the site, with concentrations ranging from 2230 to 18,000 mg/kg. The site is currently protected and monitored by the Environment Protection Authority (EPA) Victoria. Based on the environmental audit report, remediation actions are required to protect the site [52].

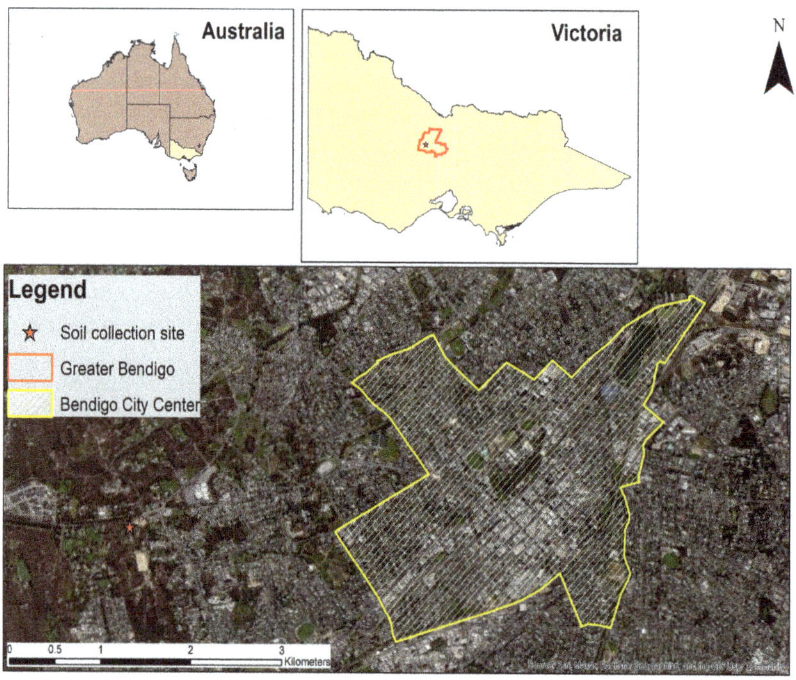

Figure 1. Map of the soil sampling site.

Mine waste soil (approximately 30 kg) from 0 to 50 cm in soil depth was collected and sieved onsite using a 2 mm mesh sieve, placed in sterile buckets with lids, and transported to the laboratory. The soil samples were stored in a 4 °C room prior to use.

2.2. Soil Properties Analysis

Soil physicochemical analysis was analysed in triplicate. Soil pH and EC were assessed using a ratio of 1:10 of dry weight soil to Milli-Q deionised water. Soil pH was measured using a HANNA Instruments edge^{pH}, and a Compact Conductivity Meter (LAQUAtwin–EC–11, HORIBA Scientific, Kyoto, Japan) was utilised to measure the EC [53]. Soil texture was determined using RemScan (Ziltek Pty. Ltd., Edwardstown, SA, Australia) [54–56]. Soil C, H, and N elements were investigated using an elemental analyser, CHN 2400 series II Perkin Elmer. The results of the analyses are presented in Table 1.

Table 1. Physicochemical properties of the soil collected from the mine site. Values represent mean and ± standard deviation (n = 3).

Properties	Value
Sand (%)	38.33 ± 3.78
Silt (%)	42.66 ± 1.15
Clay (%)	19 ± 2.64
Soil texture	Clay loam
pH	8.26 ± 0.00
Electrical conductivity (EC) (µS/cm)	2.81 ± 0.10
Moisture (%)	14.08 ± 0.11
Organic matter (%)	0.50 ± 0.01
Carbon (%)	0.28 ± 0.02
Nitrogen (%)	0.28 ± 0.02
Arsenic (mg/kg)	13,032 ± 1852.30

2.3. Plant and Biochar Collection

This study used three plants: *Juncus pauciflorus*, *Poa labillardieri*, and *Rytidosperma caespitosum*. As described in Section 1, *Poa labillardieri* and *Rytidosperma caespitosum* were selected based on previous studies. *Poa labillardieri* has been reported to accumulate As (~100 mg/kg plant biomass) [57], while *Rytidosperma caespitosum* accumulated another heavy metal, Pb (130 mg/kg). *Juncus* is a native plant growing at the soil sampling site and has also previously been investigated for As accumulation (7.26 µg/g) [49]. Tube stock plants exhibiting the same growth were collected from Edendale Nursery in Eltham, Victoria, Australia. Biochar was collected from the ARC Training Centre for the Transformation of Australia's Biosolids Resource, RMIT University; biochar was produced using biosolids obtained from the Mount Martha Municipal Wastewater recycling plant, South East Water Corporation, Melbourne, Australia, via pyrolysis at 600 °C. The properties of the biosolids biochar were previously analysed as displayed in Table 2.

2.4. Experimental Design

The Taguchi method was utilised in designing the experiment. The Taguchi method is an approach based on fractional factorial design that uses an orthogonal array to evaluate the most significant factor during treatment (Table 3) [57,58]. This study used 9 treatments with 5 replicates, as shown in Table 3. Each plant with different biochar doses was placed in pots.

Table 2. Physicochemical analyses of the biochar.

Properties	Value
pH	11.7
Electrical conductivity (EC) (μS/cm)	1100
Ammonia (mg N/kg)	170
Nitrate (mg/kg)	<0.2
Arsenic (mg/kg)	<5
BET surface area (m^2/g)	11.68

Sources: South East Water; ARC Training Centre for the Transformation of Australia's Biosolids Resource.

Table 3. Design of the experiments using the Taguchi method.

Pot	Plant Species	Biochar Concentration (*w*/*w*)
1–5	*Rytidosperma caespitosum*	0%
6–10	*Rytidosperma caespitosum*	5%
11–15	*Rytidosperma caespitosum*	10%
16–20	*Poa labillardieri*	0%
21–25	*Poa labillardieri*	5%
26–30	*Poa labillardieri*	10%
31–35	*Juncus pauciflorus*	0%
36–40	*Juncus pauciflorus*	5%
41–45	*Juncus pauciflorus*	10%

2.5. Greenhouse Experiment

Each plastic pot (1 L) with saucers was filled with sieved As-contaminated soils (500 g), following, where appropriate, thorough mixing with biochar (5 or 10% *w*/*w*). A single plant from tube stocks was transplanted into each pot and incubated for 100 days in the greenhouse before being harvested. The temperature was recorded using a temperature data logger, and plants were watered regularly to keep the moisture approximately at 80% of the soil water holding capacity.

2.6. Estimation of Plant Growth

After 100 days, plants were harvested by washing and separating shoots and roots and drying them in an oven at 60 °C for 72 h until the weight remained constant to obtain the dry weight biomass [59]. Moreover, the length of the shoots and roots were measured using a ruler in centimetres (cm).

2.7. Arsenic Analysis in Soil and Plants

Arsenic concentrations in soil and plants were analysed using ICP-MS (an Agilent 7700c with a Micromist nebuliser and X-lens). In preparation for ICP-MS, acid digestion for shoots, roots, and soil was prepared using 70% nitric acid (HNO_3) and 30% hydrogen peroxide (H_2O_2) [60]. Briefly, 0.1 g of dried plants (shoot and root) was placed in glass tubes, and 10 mL of HNO_3 was added in the tubes and placed at a heating block at 95 °C ± 5 °C and refluxed for 10–15 min. Following cooling, 5 mL of HNO_3 was added to the tubes and placed in a heating block at 95 °C ± 5 °C for 3 h. After cooling, Milli-Q water (2 mL) was added together with H_2O_2 (3 mL) and heated until the effervescence subsided; the tubes were then heated to 95 °C ± 5 °C for 2 h. The solution was transferred to 50 mL tubes and diluted with Mili-Q water, then filtered using a 0.45 µm nylon syringe filter (diameter 25 mm; FilterBio® Nantong, China), and analysed by Inductively Coupled Plasma-Mass Spectrometry (ICP-MS) using US EPA method 6020B [61].

For soil, dried soil (1 g) was placed in glass tubes, and HNO_3 (10 mL, 70%) was added, then covered with a glass marble and heated at 95 °C until no further brown fumes were released. Water (2 mL) and H_2O_2 (3 mL, 30%) were added, and the mixture was diluted to 50 mL with Mili-Q water before being filtered through a 0.45 µm filter for ICP-MS analysis. Standard reference material and blank samples were used for quality assurance and control. Montana soil (NIST2710) was used as a soil reference material, while peach leaves (NIST1547) were utilised as a plant reference.

2.8. Bioaccumulation Factor (BF) and Translocation Factor (TF) Calculations

To provide information on the efficiency of plants in translocating metals from roots to shoots, both the Bioaccumulation Factor, used to assess the roots–soil bioaccumulation [62,63] (BF) Equation (1), and the Translocation Factor, which assesses the efficiency of plants to translocate metals from roots to shoots (TF), were calculated Equation (2).

$$\mathrm{BF} = \frac{Concentration\ of\ As\ in\ shoots\ or\ roots\ (\mathrm{mg/kg})}{Concentration\ of\ As\ in\ soil\ (\mathrm{mg/kg})} \tag{1}$$

$$\mathrm{TF} = \frac{Concentration\ of\ As\ in\ shoots\ (\mathrm{mg/kg})}{Concentration\ of\ As\ in\ roots\ (\mathrm{mg/kg})} \tag{2}$$

2.9. Soil Bacterial Analysis

2.9.1. Soil DNA Extraction

Genomic DNA was extracted from soil samples with a PowerSoil® DNA Isolation Kit (MO BIO Laboratories, Inc., San Diego, CA, USA) following the manufacturer's instructions. A NanoDrop Lite spectrophotometer (Thermo Scientific, Waltham, MA, USA) was used to assess the quality and quantity of the extracted DNA. DNA samples were kept at 20 °C until further analysis.

2.9.2. 16S rRNA Analysis Using Quantitative Polymerase Chain Reaction (qPCR)

The abundance of genes coding for the 16S rRNA gene for all bacteria was measured using qPCR on a QIAGEN Rotor-Gene; each sample was run in triplicate. The combination of primers 341-F (5′CCTACGGGAGGCAGCAG3′) and 518-R (5′ATTACCGCGGCTGCTGG30) [55] was utilised for quantifying the overall bacterial levels. A reaction of 20 µL was prepared with 10 µL of qPCR mastermix (SensiFAST SYBR No-ROX Kit, Bioline, Heidelberg, Germany), 8.2 µL PCR-grade water, 1 µL DNA template, and 0.4 µL each of the forward and reverse primers. The protocol for amplifying the total bacteria included a denaturation step at 95 °C for 5 min, followed by 40 cycles of denaturation at 95 °C for 10 s, annealing at 55 °C for 30 s, extending at 72 °C for 30 s, and, finally, acquiring signals at 80 °C for 10 s [64].

2.9.3. *NifH* Gene Abundance Analysis Using Quantitative Polymerase Chain Reaction (qPCR)

To assess whether nitrogen-fixing bacteria were present in both the rhizosphere and bulk soil, the *nifH* gene was used. Two primers sets, nifHF (5′TACGGNAARGGSGGNATC-GGCAA3′) and nifHIR (5′AGCATGTCYTCSAGYTCNTCCA3′) [65], were amplified. The amplification conditions applied were an initial denaturation step at 94 °C (2 min), followed by 25 cycles of 94 °C denaturation (30 s), annealing at 60 °C (30 s), 72 °C extension (30 s), and 72 °C extension (7 min).

2.10. Statistical Analysis

Taguchi was used to determine the most affecting factor among the treatments [66,67]. The Taguchi method is used to arrange experimental designs and predict the most affecting factors. It is more beneficial than a full factorial design, as it can reduce treatment numbers, time, and cost. Taguchi uses the ratio of signal/noise (S/N), where the signal represents mean values while the noise represents the standard deviation.

$$\frac{S}{N} = 10 \times \log\left(\frac{1/Y^2}{n}\right)$$

Y represents the responses; n represents the experimental run number.

There are three categories to determine the ratio: larger is better, smaller is better, and nominal is better. In this study, we used larger is better to assess the As uptake. We used Taguchi to investigate the factors most affecting arsenic phytoremediation. Taguchi has been applied in some phytoremediation studies [58,68]. For example, Shah [2] used Taguchi to investigate the impact of banana peel biochar on cadmium-contaminated soil. Two factors (biochar and metal concentration) were designed, and they analysed the most affecting factor using the Taguchi method. In this study, we used Taguchi to investigate the most affecting factors for arsenic phytoremediation.

Before analyses, the data were tested for normality and homogeneity. ANOVA was used to assess the significance of all treatments, and Tukey's test at $p < 0.05$ was used to investigate significant differences between treatments for each plant. All analyses were conducted using Minitab-21 Software (Minitab, State College, PA, USA). Microsoft Excel and OriginPro were employed to perform the analyses.

3. Results and Discussion

3.1. Growth of Plants in Arsenic-Contaminated Soils Amended with Biosolids-Derived Biochar

All three plants were able to grow in the As-contaminated soils over the 100 days (Figure 2 and Table 4). For each plant, the final root and shoot biomass increased two-fold with elevated biochar for most of the treatments. *Juncus pauciflorus* produced the highest biomass, followed by *Poa labillardieri* and *Rytidosperma caespitosum*. *Juncus pauciflorus* in soils amended with 10% biochar had the highest root dry biomass, while *R. caespitosum* had the lowest biomass (Figure 2). Similarly, shoot biomass was significantly increased in biochar-amended soils (5 and 10%) compared to the 0% biochar treatment. The results align with a previous work reported by Guo, Chen [33], which reported an increase in *Pteris vittata* biomass following the addition of 5% biochar in comparison with 0% and 2% biochar treatments. Lin, Gao [60] also stated that biochar improved rice root biomass in soil amended with 2% biochar in As-contaminated paddy soils. Another species of *Juncus*, *Juncus subsecundus*, has been utilised to remediate Cd-contaminated soils using biochar from different sources (oil mallee plants or wheat chaff). However, biochar did not significantly enhance plant growth [69]. In this study, plant growth was also assessed using shoot and root length (Table 4). All plants with higher doses of biochar showed increased shoot and root length; however, shoot length was greater than root length. The shoot and root of *Juncus* showed the highest lengths in soil amended with 10% biochar.

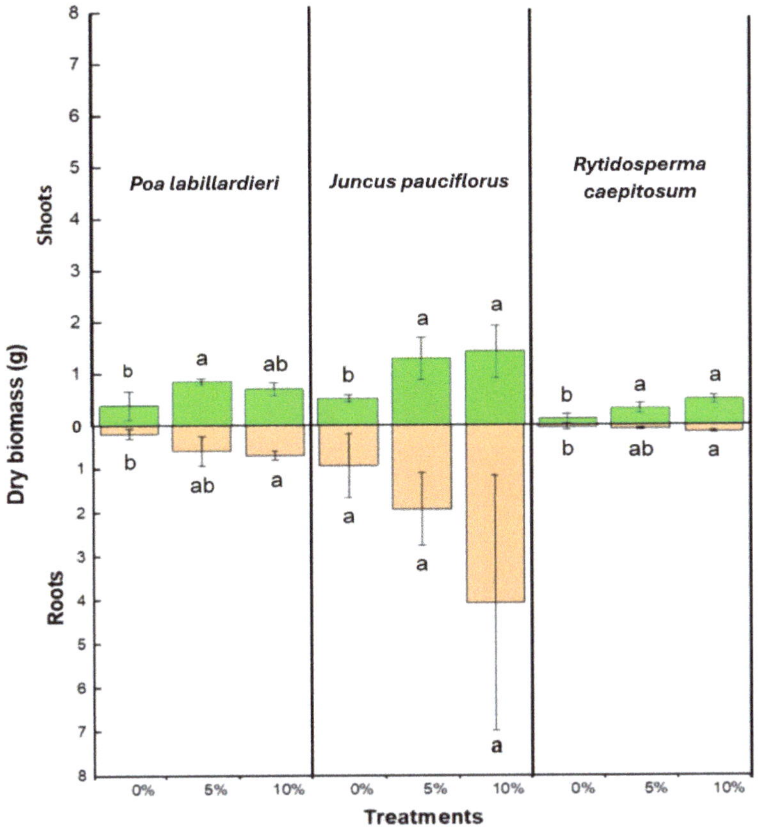

Figure 2. Shoots and roots dry biomass following growth for 100 days in As-contaminated soils with biochar addition at different doses (0%, 5%, and 10%). Different letters above the bars indicate significant differences at $p < 0.05$ (Tukey's test). Data are presented as the mean ($n = 5$). Error bars represent the standard deviation.

Table 4. Plant lengths (shoots and roots) with different biochar doses (0%, 5%, and 10%). Values represent the mean ± standard deviation ($n = 5$).

Plant Species	Biochar Dose (%)	Length (cm)	
		Shoots	Roots
Poa labillardieri	0	23.1 ± 2.81	13 ± 3.51
Poa labillardieri	5	28 ± 8.82	17.7 ± 6.90
Poa labillardieri	10	29 ± 6.28	19.5 ± 7.31
Juncus pauciflorus	0	22 ± 3.24	16.2 ± 2.41
Juncus pauciflorus	5	26 ± 4.28	16 ± 3.53
Juncus pauciflorus	10	30.9 ± 5.67	19.68 ± 4.68
Rytidosperma caespitosum	0	17.5 ± 3.37	7.4 ± 1.94
Rytidosperma caespitosum	5	25.7 ± 1.75	11.7 ± 2.51
Rytidosperma caespitosum	10	27.6 ± 2.90	13.8 ± 4.39

3.2. The Effect of Biochar on Arsenic Uptake in Plants

Figure 3 shows the results from the assessment of the impact of the biochar amendment of As-contaminated mine waste soils on plant As uptake. Generally, for all plants, the As concentration (per unit biomass) in both shoots and roots was higher in plants in soil with no biochar addition, with the exception of *J. pauciflorus* shoot and *R. caespitosum* root, in which the highest As concentrations were found in mine waste soils amended with 5% biochar (Figure 3a). In terms of roots, the highest As concentration was recorded in *J. pauciflorus* with 0% biochar (5327.82 mg/kg), while *P. labillardieri* showed the lowest As concentration (819.45 mg/kg) in soils amended with 10% biochar. For shoots, *P. labillardieri* in soils without biochar had an As concentration of 1202.72 mg/kg, significantly higher than shoots from the same plant in soils amended with 5% and 10% biochar treatments (Figure 3a).

(a)

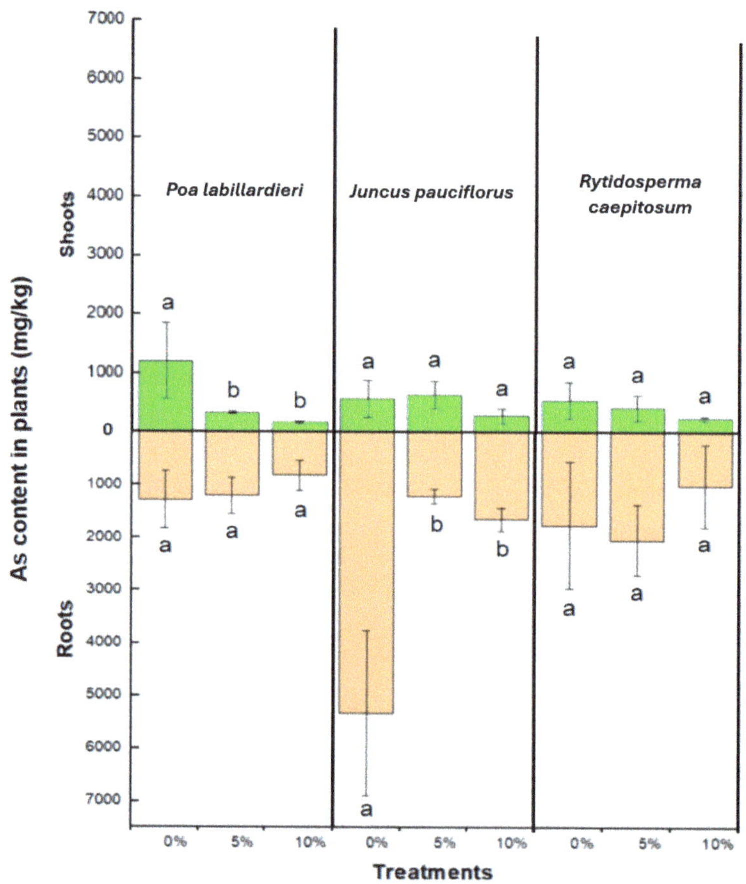

Figure 3. *Cont.*

(b)

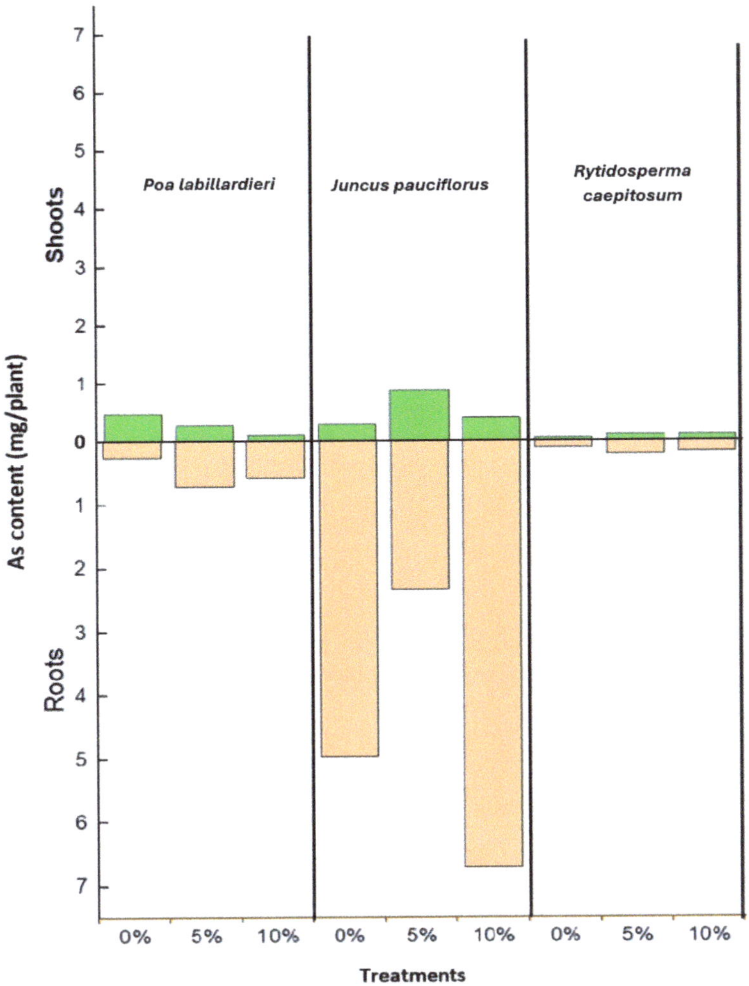

Figure 3. (**a**) Arsenic accumulation in shoots and roots following biochar addition with different doses (0%, 5%, and 10%). Different letters above the bars indicate significant differences at $p < 0.05$ (Tukey's test). Data are presented as the mean ($n = 5$). Error bars represent the standard deviation. (**b**) Total arsenic uptake per plant (shoots and roots) at different biochar doses (0%, 5%, and 10%).

Figure 3b illustrates the total roots and shoots uptake of As per plant for each of the three plants, illustrating the impact of the increased growth of plants in biochar and the reduced As uptake per unit biomass in soils amended with biochar. This study reveals that *J. pauciflorus* accumulated significantly more As in comparison to the other plants, with soil amended with 10% biochar leading to more As uptake per plant (7.10 mg), 35% greater than As accumulated by *J. pauciflorus* in unamended mine waste soils (5.25 mg). The uptake of As in the other two species were all below 1 mg, independent of the treatment (Figure 3b). In this study, the higher dose of biochar increased the biomass of all the plants but reduced the As uptake per unit biomass. A previous study performed by Zheng, Wang [70] also showed a reduction in As uptake by the roots of *Pteris vittata* following 3% biochar application compared to 1% and 2% biochar. The high rate of biochar might cause an excess in As mobilisation that reduces the capability of the plants to uptake As; therefore, it is vital to determine an appropriate incorporation rate of biochar in the phytoremediation implementation.

3.3. *Effect of Biochar on Arsenic Concentration in Soil*

A comparison of the As concentration pre- and post-remediation is shown in Figure 4 and confirms a general reduction in soil As concentration in soils that underwent phytoremediation. Across all plants, the As concentration in soils amended with 10% biochar was significantly lower. However, the As concentrations in all the treatments were not significantly different, perhaps reflecting some heterogeneity within the mine waste soils collected from a contaminated site. These results align with a study by Liu, Yang [38] that reported a decline in As concentration in mine waste soil after the addition of 1.5% and 4% compared to 0% and 5%, from ±115 mg/kg to ±105 mg/kg. However, no significant differences were observed between the soil As concentrations in soils amended with 1.5% or 4% biochar. A similar result was also reported by Guo, Chen [33], who used 0%, 2%, and 5% biochar to remediate As-contaminated soil using *Pteris vittata*. Biochar applied at a higher rate reduced the As concentration in soil in comparison with the control. The soil pH after biochar treatments ranged between 7.5 and 8.5, which did not significantly change compared to the initial pH (8.26).

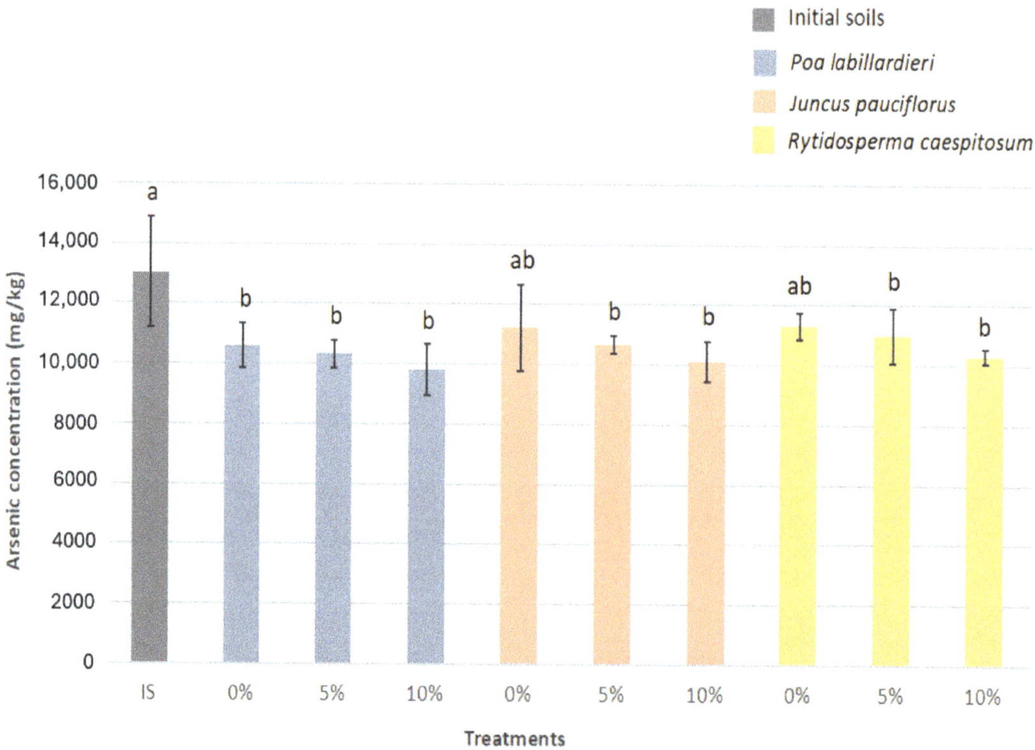

Figure 4. Arsenic concentration in soils before and after biochar addition (0%, 5%, and 10%). Different letters above the bars indicate significant differences at $p < 0.05$ (Tukey's test). Data are presented as the mean ($n = 5$). Error bars represent the standard deviation.

3.4. Translocation Factor (TF) and Bioaccumulation Factor (BAF) for Treated Plants

The Translocation Factor (TF) was assessed to identify the potential of plants for phytoremediation. If plants exhibit BAF and TF < 1, they are potentially useful for phytostabilisation (Table 5). However, plants with BAF and TF > 1 indicate their ability for phytoextraction [63]. This study reveals that *J. pauciflorus* and *R. caespitosum* are suitable for phytostabilisation, as they both have coefficient factors < 1. In contrast, *P. labillardieri* shows a TF > 1 (Table 5). Although the results confirm that *P. labillardieri* is capable of phytoextraction, relatively low levels of biomass were produced (Figure 2) in these soils, even with biochar. *Juncus pauciflorus* has the highest BAF coefficient in roots (0.45 ± 0.12) without biochar addition, which, along with the fact that this plant produced the greatest roots biomass (Figure 2), indicates that the root has more As accumulation, making it the most promising plant in terms of phytostabilisation. The results show that the addition of biochar decreases the BAF and TF in both shoots and roots. A study by Jing, Yang [71] showed a similar result, as both TF and BAF decreased in rice tissues with a higher rate of biochar application (ranging from 10 to 40 t/ha) in As-contaminated soils. Mohamed, Zhang [72] stated that this decrease can be caused by the strong capacity of biochar to absorb As.

Table 5. Bioaccumulation Factor (BAF) and Translocation Factor (TF) for treated plants with biochar. Data are presented as the mean ± standard deviation (n = 5).

Plant Species	Biochar Dose (%)	BAF in Shoots	BAF in Roots	TF
Poa labillardieri	0	0.11 ± 0.06	0.12 ± 0.05	1.06 ± 0.82
	5	0.03 ± 0.00	0.11 ± 0.02	0.29 ± 0.12
	10	0.02 ± 0.00	0.08 ± 0.03	0.22 ± 0.09
Juncus pauciflorus	0	0.05 ± 0.03	0.45 ± 0.12	0.11 ± 0.05
	5	0.06 ± 0.02	0.11 ± 0.01	0.57 ± 0.21
	10	0.03 ± 0.01	0.16 ± 0.02	0.17 ± 0.07
Rytidosperma caespitosum	0	0.05 ± 0.03	0.15 ± 0.10	0.61 ± 0.81
	5	0.04 ± 0.02	0.18 ± 0.06	0.20 ± 0.08
	10	0.02 ± 0.00	0.10 ± 0.08	0.40 ± 0.38

3.5. Taguchi Analysis for Arsenic Uptake

To compare the relative importance of the experimental factors tested, Taguchi analysis was conducted. The results confirmed that both factors, plant species and biochar concentration, affected As uptake (Figure 5). Plant species occupied the first rank (Delta: 4.9347), confirming that plant species have the largest effect in terms of As uptake, followed by biochar as the second rank. Level 2 in Table 6 belongs to *J. pauciflorus*, confirming that this plant shows significantly higher As uptake than the other plants. Interestingly, *Juncus* is a genus found growing at the soil sampling site. Its ability to grow in As-contaminated soils might confirm its potential suitability for phytoremediation. As suggested by Barbafieri, Dadea [73], native plants in contaminated tailings can tolerate elevated metal concentrations. Therefore, *J. pauciflorus* may be more suited to the local soil and environmental conditions compared to the other plants. In terms of the second rank, the 10% biochar concentration showed the optimum treatment, followed by no biochar amendment (Table 6). To the best of the authors' knowledge, there are no previous studies of As phytoremediation that used the Taguchi method to investigate the treatment factors. However, Taguchi was used by Shah, Soni [58] to analyse the comparison of two factors: biochar concentration and Cd concentration on the phytoremediation of Cd in soils by *Bidens pilosa*. They found that biochar concentration was a more important factor than Cd concentration during the phytoremediation.

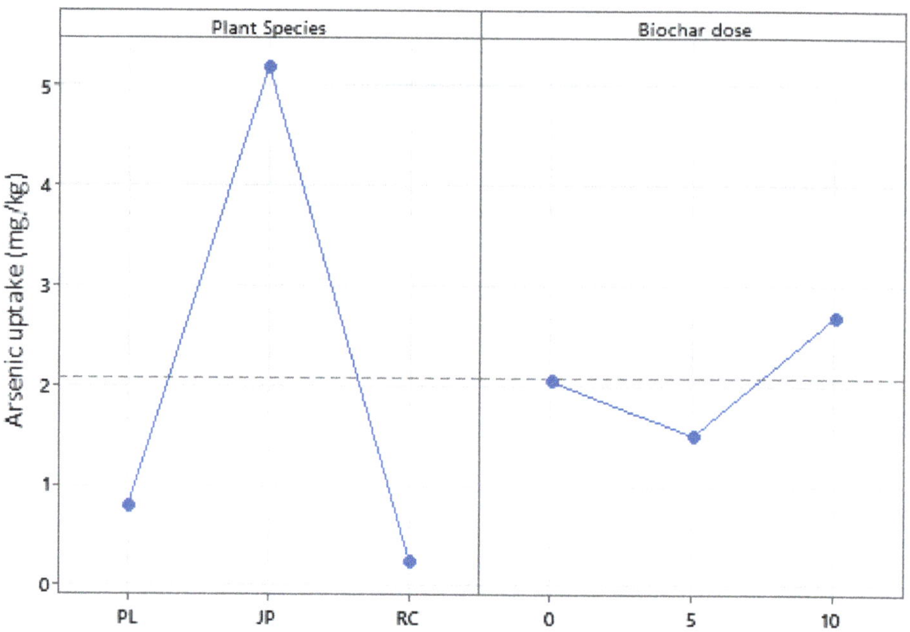

Figure 5. Plot of the mean effect of the experimental factors. PL: *Poa labillardieri*; JP: *Juncus pauciflorus*; RC: *Rytidosperma caespitosum*. Biochar dose (0%, 5%, and 10%).

Table 6. Response table for means from the Taguchi analysis of the experimental data. Delta is the range of average response values between the highest and lowest for each factor. Rank 1 belongs to the highest Delta value, and Rank 2 is the second highest.

Level	Plant Species	Biochar Dose
1	0.8002	2.0463
2	5.1885	1.5044
3	0.2537	2.6917
Delta	4.9347	1.1873
Rank	1	2

3.6. Effect of Biochar on Soil Microbial Community

Two microbial parameters, copy numbers of 16S rRNA and *nifH* genes, were assessed to determine whether biochar affected the abundance of the total bacterial community and nitrogen-fixing bacteria, respectively (Figure 6a,b). The presence of nitrogen-fixing bacteria is vital, because they can provide nitrogen for soil through nitrogen fixation, and the nitrogen can be absorbed by plants to grow [74,75].

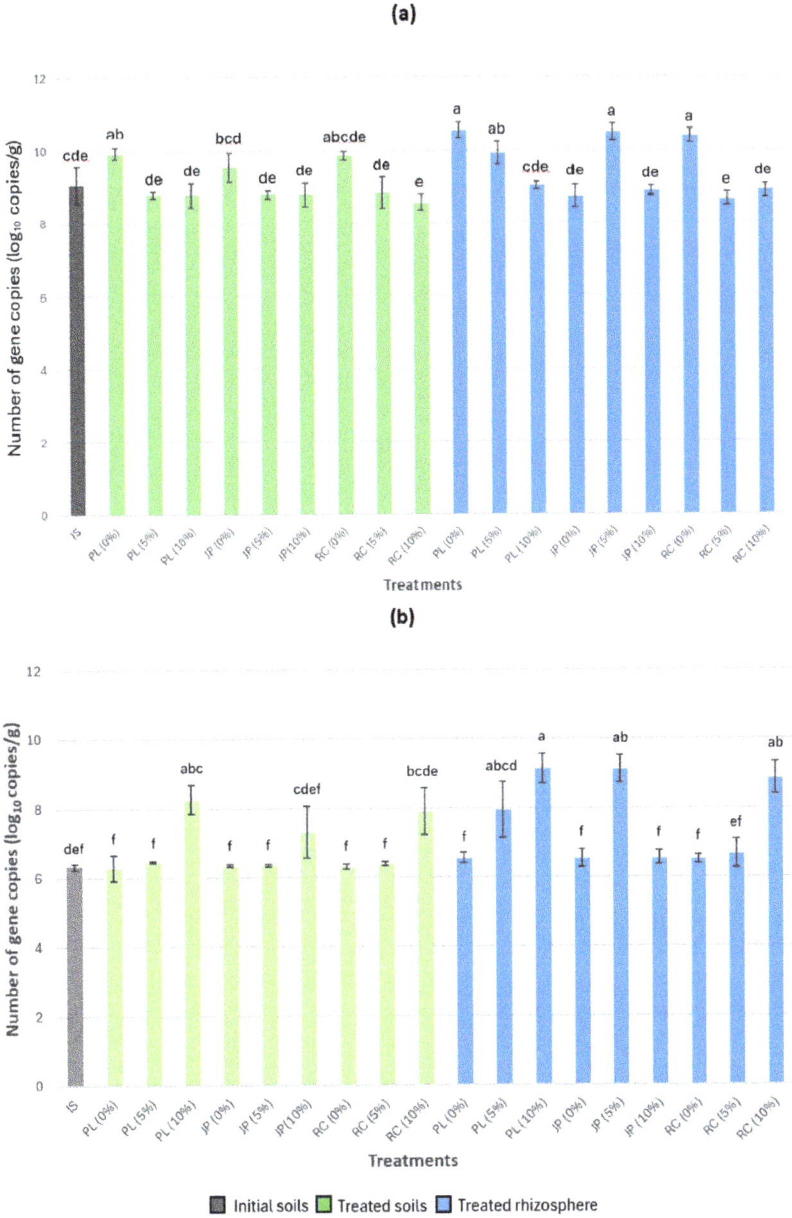

Figure 6. (**a**) Variation in the number of gene copies (16S rRNA) (log_{10} copies/g) measured using qPCR for the initial soils, treated soils, and rhizosphere after biochar addition (0%, 5%, and 10%). PL: *P. labillardieri*; JP: *J. pauciflorus*; RC: *R. caespitosum*. Different letters above the bars indicate significant differences at $p < 0.05$ (Tukey's test). Data are presented as the mean ($n = 3$). Error bars represent the standard deviation. (**b**) Variation in the number of gene copies (*nifH*) (log_{10} copies/g) measured using qPCR for the initial soils, treated soils, and rhizosphere after biochar addition (0%, 5%, and 10%). PL: *P. labillardieri*; JP: *J. pauciflorus*; RC: *R. caespitosum*. Different letters above the bars indicate significant differences at $p < 0.05$ (Tukey's test). Data are presented as the mean ($n = 3$). Error bars represent the standard deviation.

Soil Microbial Abundance of Total Bacteria and Nitrogen-Fixing Bacteria in Bulk Soil and Rhizosphere

The abundance of the 16S rRNA gene indicates the total bacteria in the untreated and treated soils (Figure 6a). The results reveal that 5% and 10% biochar reduce the bacterial number of gene copies in treated soil (bulk soil) and rhizosphere soil of *P. labillardieri* and *R. caespitosum* but not in *J. pauciflorus* (Figure 6a). The highest number of gene copies was found in the rhizosphere of *P. labillardieri* with 0% biochar (10.53 log_{10} copies/g), followed by *J. pauciflorus* with 5% biochar (10.48 log_{10} copies/g) and *R. caespitosum* with 0% biochar (10.38 log_{10} copies/g). Several studies have shown that the addition of biochar can impact the soil microbial community in contaminated soils; however, few studies have focused on As phytoremediation. Shang, Wu [76] demonstrated that the application of rice husk biochar for PAH phytoremediation resulted in a significant increase in the number of 16S rRNA gene copies in the amended soil. The current study demonstrated that biochar addition led to a reduction in the number of 16S gene copies. This may also explain why the uptake of As by roots grown in biochar-amended soil was lower per unit plant biomass than taken up by roots of plants in unamended soils. Huang [77] stated that As uptake is influenced by soil microbes and involves several mechanisms. Arsenic can be sorbed by the bacterial cell surface and is dependent on pH ranging from 4 to 7.5 [77]. After sorption by the microbial cell surface, As then accumulates in cell membranes and cytoplasm. The reduction of As can occur through intracellular pathways [78]. Although As-resistant microbes are widely distributed in the environment, not many are As hyperaccumulators [78].

Unlike the 16S rRNA gene, a significantly higher number of gene copies of the *nifH* gene was associated with bulk and rhizosphere soils amended with 10% biochar, except for *J. pauciflorus*, which had the highest number of gene copies in soils amended with 5% biochar. The highest number of *nifH* gene was found in *P. labillardieri* (9.13 log_{10} copies/g); the lowest gene copies (6.51 log_{10} copies/g) were found in *R. caespitosum* with 0% of biochar. The biochar used in this study contains nitrogen (0.28%) (Table 1), which has the potential to enhance plant growth. The presence of *nifH* gene is important in showing the availability of diazotrophs that might provide nitrogen for plant growth.

To date, limited studies have been conducted to assess the impact of biochar on *nifH* gene copies in soil during As phytoremediation. One study by Zhang, Bai [79] showed that, at a higher biochar concentration (6%), a reduction in the abundance of the *nifH* gene from 1.8×10^7 copies/g to 1.3×10^7 copies/g occurred during the phytoremediation of agricultural soil contaminated with heavy metals using *Medicago sativa* L. (alfalfa). The authors speculated that the increased soil pH following biochar application may explain the reduction in the number of nitrogen-fixing organisms. Soil pH is one of the key environmental factors that significantly affect the retention of ions and the availability of nutrients. In the current study, the increased abundance of the *nifH* gene in soils amended with biochar, particularly at 10%, was positively correlated with the plant roots and shoots biomass. To date, the impact of biochar on this important bacterial group has not been studied, although the impact of biochar on increasing plant biomass production has been studied [74].

4. Conclusions

This study showed that As-contaminated mine waste soils amended with *J. pauciflorus*, together with biochar (10%), resulted in the highest increase of plant biomass and As uptake per plant as compared to the other treatments. Examination of the impact of the amendment on the soil microbial community revealed that biochar addition resulted in a reduction in the total bacterial community, while the population of nitrogen-fixing bacteria, as shown by the number of gene copies of the *nifH* gene, significantly increased with the increasing biochar concentration. Future research is required to further examine whether the utilisation of other soil amendments, such as nitrogen-fixing bacteria, would enhance As uptake with or without biochar addition. This study confirmed, for the first time, that biosolids-derived biochar enhanced the phytoremediation of As-contaminated soils, and further work must be conducted to compare the properties of the biochar pre- and post-treatment; this work could lead to a better understanding of the role of biochar in phytoremediation. Moreover, pilot-scale experiments should be conducted to investigate the ability of biochar and other amendments to improve As phytoremediation under field conditions.

Author Contributions: Conceptualisation, F.H., L.S.K., K.S., A.S., P.N. and A.S.B.; investigation, F.H.; resources, A.S.B.; data curation, F.H.; writing—original draft preparation, F.H.; writing—review and editing, L.S.K., K.N., J.A.B., K.S., A.S., P.N. and A.S.B.; visualisation, F.H., L.S.K., K.N., K.S., A.S., P.N. and A.S.B.; supervision, L.S.K., K.S., J.A.B., K.N., A.S., P.N. and A.S.B.; project administration, F.H. and A.S.B.; funding acquisition, A.S.B. All authors have read and agreed to the published version of the manuscript.

Funding: This research was supported by the RMIT University research scholarship and Beasiswa Kementerian Agama—LPDP (Scholarship of Ministry of Religious Affair Republic of Indonesia—Indonesia Endowment Fund for Education).

Data Availability Statement: The original contributions presented in this study are included in the article. Further inquiries can be directed to the corresponding author(s).

Acknowledgments: The authors acknowledge RMIT University, ARC Training Centre for the Transformation of Australia's Biosolids Resource, and Beasiswa Kementerian Agama—LPDP (Scholarship of Ministry of Religious Affair Republic of Indonesia—Indonesia Endowment Fund for Education) for supporting this research.

Conflicts of Interest: The authors declare no conflicts of interest.

References

1. Khan, S.; Naushad, M.; Lima, E.C.; Zhang, S.; Shaheen, S.M.; Rinklebe, J. Global soil pollution by toxic elements: Current status and future perspectives on the risk assessment and remediation strategies–A review. *J. Hazard. Mater.* **2021**, *417*, 126039. [PubMed]
2. Preetha, J.S.Y.; Arun, M.; Vidya, N.; Kowsalya, K.; Halka, J.; Ondrasek, G. Biotechnology advances in bioremediation of arsenic: A review. *Molecules* **2023**, *28*, 1474. [CrossRef] [PubMed]
3. ATSDR's Substance Priority List. Available online: https://www.atsdr.cdc.gov/SPL/# (accessed on 24 March 2024).
4. Chowdhury, U.K.; Biswas, B.K.; Chowdhury, T.R.; Samanta, G.; Mandal, B.K.; Basu, G.C.; Chanda, C.R.; Lodh, D.; Saha, K.C.; Mukherjee, S.K. Groundwater arsenic contamination in Bangladesh and West Bengal, India. *Environ. Health Perspect.* **2000**, *108*, 393–397. [PubMed]
5. Sun, G.; Dai, G.; Li, F.; Yamauchi, H.; Yoshida, T.; Aikawa, H. The present situation of chronic arsenism and research in China. In *Arsenic Exposure and Health Effects*; Elsevier: Amsterdam, The Netherlands, 1999; pp. 123–126.
6. Tseng, W.; Chu, H.M.; How, S.; Fong, J.; Lin, C.; Yeh, S. Prevalence of skin cancer in an endemic area of chronic arsenicism in Taiwan. *J. Natl. Cancer Inst.* **1968**, *40*, 453–463.
7. Chiou, H.-Y.; Huang, W.-I.; Su, C.-L.; Chang, S.-F.; Hsu, Y.-H.; Chen, C.-J. Dose-response relationship between prevalence of cerebrovascular disease and ingested inorganic arsenic. *Stroke* **1997**, *28*, 1717–1723. [CrossRef]
8. Kumar, A.; Ali, M.; Kumar, R.; Kumar, M.; Sagar, P.; Pandey, R.K.; Akhouri, V.; Kumar, V.; Anand, G.; Niraj, P.K. Arsenic exposure in Indo Gangetic plains of Bihar causing increased cancer risk. *Sci. Rep.* **2021**, *11*, 2376. [CrossRef]
9. Jang, Y.; Somanna, Y.; Kim, H. Source, distribution, toxicity and remediation of arsenic in the environment—A review. *Int. J. Appl. Environ. Sci.* **2016**, *11*, 559–581.
10. Murcott, S. *Arsenic Contamination in the World*; IWA Publishing: London, UK, 2012.
11. de Souza Neto, H.F.; da Silveira Pereira, W.V.; Dias, Y.N.; de Souza, E.S.; Teixeira, R.A.; de Lima, M.W.; Ramos, S.J.; do Amarante, C.B.; Fernandes, A.R. Environmental and human health risks of arsenic in gold mining areas in the eastern Amazon. *Environ. Pollut.* **2020**, *265*, 114969. [CrossRef]
12. Yang, F.; Xie, S.; Wei, C.; Liu, J.; Zhang, H.; Chen, T.; Zhang, J. Arsenic characteristics in the terrestrial environment in the vicinity of the Shimen realgar mine, China. *Sci. Total Environ.* **2018**, *626*, 77–86. [CrossRef]

13. Smith, E.; Smith, J.; Smith, L.; Biswas, T.; Correll, R.; Naidu, R. Arsenic in Australian environment: An overview. *J. Environ. Sci. Health Part A* **2003**, *38*, 223–239.
14. Álvarez-Quintana, J.; Álvarez, R.; Ordóñez, A. Arsenic in soils affected by mining: Microscopic studies vs. sequential chemical extraction. *Int. J. Environ. Res. Public Health* **2020**, *17*, 8426. [CrossRef] [PubMed]
15. Eisler, R. Mercury hazards from gold mining to humans, plants, and animals. In *Reviews of Environmental Contamination and Toxicology*; Springer: New York, NY, USA, 2004; pp. 139–198.
16. Ko, I.; Sung Ahn, J.; Seog Park, Y.; Kim, K.-W. Arsenic contamination of soils and sediments from tailings in the vicinity of Myungbong Au mine, Korea. *Chem. Speciat. Bioavailab.* **2003**, *15*, 67–74.
17. Williams, M. Arsenic in mine waters: An international study. *Environ. Geol.* **2001**, *40*, 267–278.
18. Liao, X.; Li, Y.; Miranda-Avilés, R.; Zha, X.; Anguiano, J.H.H.; Sánchez, C.D.M.; Puy-Alquiza, M.J.; González, V.P.; Garzon, L.F.R. In situ remediation and ex situ treatment practices of arsenic-contaminated soil: An overview on recent advances. *J. Hazard. Mater. Adv.* **2022**, *8*, 100157.
19. Huslina, F.; Khudur, L.S.; Shah, K.; Surapaneni, A.; Netherway, P.; Ball, A.S. Mine Site Restoration: The Phytoremediation of Arsenic-Contaminated Soils. *Environments* **2024**, *11*, 99. [CrossRef]
20. Peco, J.D.; Higueras, P.; Campos, J.A.; Esbrí, J.M.; Moreno, M.M.; Battaglia-Brunet, F.; Sandalio, L.M. Abandoned mine lands reclamation by plant remediation technologies. *Sustainability* **2021**, *13*, 6555. [CrossRef]
21. Ghori, Z.; Iftikhar, H.; Bhatti, M.F.; Sharma, I.; Kazi, A.G.; Ahmad, P. Phytoextraction: The use of plants to remove heavy metals from soil. In *Plant Metal Interaction*; Elsevier: Amsterdam, The Netherlands, 2016; pp. 385–409.
22. Islam, M.S.; Akter, R.; Rahman, M.M.; Kurasaki, M. Phytoremediation: Background, Principle, and Application, Plant Species Used for Phytoremediation. In *Design of Materials and Technologies for Environmental Remediation*; Springer: Berlin/Heidelberg, Germany, 2022; pp. 199–224.
23. Wan, X.; Lei, M.; Chen, T. Review on remediation technologies for arsenic-contaminated soil. *Front. Environ. Sci. Eng.* **2020**, *14*, 24.
24. Singh, S.; Karwadiya, J.; Srivastava, S.; Patra, P.K.; Venugopalan, V. Potential of indigenous plant species for phytoremediation of arsenic contaminated water and soil. *Ecol. Eng.* **2022**, *175*, 106476.
25. Niazi, N.K.; Singh, B.; Van Zwieten, L.; Kachenko, A.G. Phytoremediation of an arsenic-contaminated site using *Pteris vittata* L. and *Pityrogramma calomelanos* var. austroamericana: A long-term study. *Environ. Sci. Pollut. Res.* **2012**, *19*, 3506–3515.
26. Martínez-López, S.; Martínez-Sánchez, M.J.; Pérez-Sirvent, C.; Bech, J.; del Carmen Gómez Martínez, M.; García-Fernandez, A.J. Screening of wild plants for use in the phytoremediation of mining-influenced soils containing arsenic in semiarid environments. *J. Soils Sediments* **2014**, *14*, 794–809.
27. Ghosh, D.; Maiti, S.K. Biochar assisted phytoremediation and biomass disposal in heavy metal contaminated mine soils: A review. *Int. J. Phytoremediation* **2021**, *23*, 559–576. [CrossRef] [PubMed]
28. Wang, L.; Ji, B.; Hu, Y.; Liu, R.; Sun, W. A review on in situ phytoremediation of mine tailings. *Chemosphere* **2017**, *184*, 594–600. [CrossRef] [PubMed]
29. Mehmood, T.; Liu, C.; Niazi, N.K.; Gaurav, G.K.; Ashraf, A.; Bibi, I. Compost-mediated arsenic phytoremediation, health risk assessment and economic feasibility using *Zea mays* L. in contrasting textured soils. *Int. J. Phytoremediation* **2021**, *23*, 899–910. [CrossRef]
30. Yang, J.; Yang, S.-S.; Lei, M.; Yang, J.-X.; Wan, X.-M.; Chen, T.-B.; Wang, X.-L.; Guo, G.-H.; Guo, J.-M.; Liu, S.-Q. Comparison among soil additives for enhancing *Pteris vittata* L.: Phytoremediation of As-contaminated soil. *Int. J. Phytoremediation* **2018**, *20*, 1300–1306. [CrossRef] [PubMed]
31. Souri, Z.; Karimi, N.; Norouzi, L.; Ma, X. Elucidating the physiological mechanisms underlying enhanced arsenic hyperaccumulation by glutathione modified superparamagnetic iron oxide nanoparticles in *Isatis cappadocica*. *Ecotoxicol. Environ. Saf.* **2020**, *206*, 111336. [CrossRef]
32. Srivastava, S.; Shukla, A.; Rajput, V.D.; Kumar, K.; Minkina, T.; Mandzhieva, S.; Shmaraeva, A.; Suprasanna, P. Arsenic Remediation through Sustainable Phytoremediation Approaches. *Minerals* **2021**, *11*, 936. [CrossRef]
33. Guo, G.; Chen, S.; Zhang, D.; Wang, J.; Lei, M.; Ju, T.; Wei, H. Influence of biochar on the arsenic phytoextraction potential of *Pteris vittata* in soils from an abandoned arsenic mining site. *Chemosphere* **2024**, *352*, 141389. [CrossRef]
34. Vijayaraghavan, K. Recent advancements in biochar preparation, feedstocks, modification, characterization and future applications. *Environ. Technol. Rev.* **2019**, *8*, 47–64. [CrossRef]
35. Dike, C.C.; Shahsavari, E.; Surapaneni, A.; Shah, K.; Ball, A.S. Can biochar be an effective and reliable biostimulating agent for the remediation of hydrocarbon-contaminated soils? *Environ. Int.* **2021**, *154*, 106553. [CrossRef]
36. Yang, D.; Yunguo, L.; Shaobo, L.; Huang, X.; Zhongwu, L.; Xiaofei, T.; Guangming, Z.; Lu, Z. Potential benefits of biochar in agricultural soils: A review. *Pedosphere* **2017**, *27*, 645–661.
37. Varjani, S.; Kumar, G.; Rene, E.R. Developments in biochar application for pesticide remediation: Current knowledge and future research directions. *J. Environ. Manag.* **2019**, *232*, 505–513. [CrossRef] [PubMed]

38. Liu, Z.-Y.; Yang, R.; Xiang, X.-Y.; Niu, L.-L.; Yin, D.-X. Enhancement of phytoextraction efficiency coupling *Pteris vittata* with low-dose biochar in arsenic-contaminated soil. *Int. J. Phytoremediation* **2023**, *25*, 1810–1818. [CrossRef] [PubMed]
39. Gascó, G.; Álvarez, M.L.; Paz-Ferreiro, J.; Méndez, A. Combining phytoextraction by Brassica napus and biochar amendment for the remediation of a mining soil in Riotinto (Spain). *Chemosphere* **2019**, *231*, 562–570. [CrossRef] [PubMed]
40. van Dijk, L.J.; Abdelfattah, A.; Ehrlén, J.; Tack, A.J. Soil microbiomes drive aboveground plant–pathogen–insect interactions. *Oikos* **2022**, *2022*, e09366. [CrossRef]
41. Ji, H.; Zhang, Y.; Bararunyeretse, P.; Li, H. Characterization of microbial communities of soils from gold mine tailings and identification of mercury-resistant strain. *Ecotoxicol. Environ. Saf.* **2018**, *165*, 182–193. [CrossRef]
42. Sun, W.; Ji, B.; Khoso, S.A.; Tang, H.; Liu, R.; Wang, L.; Hu, Y. An extensive review on restoration technologies for mining tailings. *Environ. Sci. Pollut. Res.* **2018**, *25*, 33911–33925. [CrossRef]
43. Cross, A.T.; Ivanov, D.; Stevens, J.C.; Sadler, R.; Zhong, H.; Lambers, H.; Dixon, K.W. Nitrogen limitation and calcifuge plant strategies constrain the establishment of native vegetation on magnetite mine tailings. *Plant Soil.* **2021**, *461*, 181–201. [CrossRef]
44. Huang, L.-N.; Tang, F.-Z.; Song, Y.-S.; Wan, C.-Y.; Wang, S.-L.; Liu, W.-Q.; Shu, W.-S. Biodiversity, abundance, and activity of nitrogen-fixing bacteria during primary succession on a copper mine tailings. *FEMS Microbiol. Ecol.* **2011**, *78*, 439–450. [CrossRef]
45. Hayes, W.; Chaudhry, T.; Buckney, R.; Khan, A. Phytoaccumulation of trace metals at the Sunny Corner mine, New South Wales, with suggestions for a possible remediation strategy. *Australas. J. Ecotoxicol.* **2003**, *9*, 69–82.
46. De Silva, S.; Bernett, C.; Meaklim, J.; Abeywardane, E.; Reichman, S.M. Probing the effects of different lead compounds on the bioavailability of lead to plants. *Chemosphere* **2019**, *230*, 24–28. [CrossRef]
47. Noble, T.; Lottermoser, B. Evaluating the remediation of arsenic-rich tailings at the historic Royal George tin mine, Tasmania, Australia. In *Mine Closure 2012: Proceedings of the Seventh International Conference on Mine Closure*; Australian Centre for Geomechanics: Crawley, Australia, 2012; pp. 163–172.
48. Noble, T.; Lottermoser, B.; Townsend, A. Mobility of arsenic and environmentally significant elements in mine tailings following liming. *Aust. J. Earth Sci.* **2016**, *63*, 781–793. [CrossRef]
49. Alam, M.R.; Rahman, M.M.; Tam, N.F.-Y.; Yu, R.M.K.; MacFarlane, G.R. The accumulation and distribution of arsenic species and selected metals in the saltmarsh halophyte, spiny rush (*Juncus acutus*). *Mar. Pollut. Bull.* **2022**, *175*, 113373. [CrossRef] [PubMed]
50. Patel, S.; Kundu, S.; Halder, P.; Ratnnayake, N.; Marzbali, M.H.; Aktar, S.; Selezneva, E.; Paz-Ferreiro, J.; Surapaneni, A.; de Figueiredo, C.C. A critical literature review on biosolids to biochar: An alternative biosolids management option. *Rev. Environ. Sci. Biotechnol.* **2020**, *19*, 807–841. [CrossRef]
51. ANZBP. Australian Biosolids Statistics. Available online: https://www.biosolids.com.au/guidelines/australian-biosolids-statistics/ (accessed on 1 July 2024).
52. Austin, S.; Dobson, K. *Environmental Audit (s 53V): Gold Mining Tailings at Clay Gully, Maiden Gully, Victoria*; AECOM, NSW, Australia Pty Ltd.: Sydney, NSW, Australia, 2021.
53. Ngo, T.; Khudur, L.S.; Hakeem, I.G.; Shah, K.; Surapaneni, A.; Ball, A.S. Wood biochar enhances the valorisation of the anaerobic digestion of chicken manure. *Clean. Technol.* **2022**, *4*, 420–439. [CrossRef]
54. Thomas, C.L.; Hernandez-Allica, J.; Dunham, S.J.; McGrath, S.P.; Haefele, S.M. A comparison of soil texture measurements using mid-infrared spectroscopy (MIRS) and laser diffraction analysis (LDA) in diverse soils. *Sci. Rep.* **2021**, *11*, 16. [CrossRef]
55. Soriano-Disla, J.M.; Janik, L.J.; Forrester, S.T.; Grocke, S.F.; Fitzpatrick, R.W.; McLaughlin, M.J. The use of mid-infrared diffuse reflectance spectroscopy for acid sulfate soil analysis. *Sci. Total Environ.* **2019**, *646*, 1489–1502. [CrossRef]
56. Janik, L.J.; Soriano-Disla, J.M.; Forrester, S.T. Feasibility of handheld mid-infrared spectroscopy to predict particle size distribution: Influence of soil field condition and utilisation of existing spectral libraries. *Soil. Res.* **2020**, *58*, 528–539. [CrossRef]
57. Unal, R.; Dean, E.B. Taguchi approach to design optimization for quality and cost: An overview. In Proceedings of the 1991 Annual Conference of the International Society of Parametric Analysts, Online, 1 January 1991.
58. Shah, V.; Soni, V.; Daverey, A. Valorization of banana peel as biochar and assessment of its effect in biochar-assisted phytoremediation of cadmium-contaminated soil by using the Taguchi method. *Biomass Convers. Biorefinery* **2023**, *13*, 9451–9463. [CrossRef]
59. Chuaphasuk, C.; Prapagdee, B. Effects of biochar-immobilized bacteria on phytoremediation of cadmium-polluted soil. *Environ. Sci. Pollut. Res.* **2019**, *26*, 23679–23688. [CrossRef]
60. Cai, Y.; Georgiadis, M.; Fourqurean, J.W. Determination of arsenic in seagrass using inductively coupled plasma mass spectrometry. *Spectrochim. Acta Part B At. Spectrosc.* **2000**, *55*, 1411–1422. [CrossRef]
61. EPA. *Method 6020B Inductively Coupled Plasma-Mass Spectrometry*; EPA: Washington, DC, USA, 2014. Available online: https://www.epa.gov/sites/default/files/2015-12/documents/6020b.pdf (accessed on 4 February 2025).
62. Yoon, J.; Cao, X.; Zhou, Q.; Ma, L.Q. Accumulation of Pb, Cu, and Zn in native plants growing on a contaminated Florida site. *Sci. Total Environ.* **2006**, *368*, 456–464. [CrossRef] [PubMed]
63. Sharma, P.; Singh, S.P.; Tong, Y.W. Phytoremediation of metals: Bioconcentration and translocation factors. In *Current Developments in Biotechnology and Bioengineering*; Elsevier: Amsterdam, The Netherlands, 2022; pp. 19–37.

64. Shahsavari, E.; Aburto-Medina, A.; Taha, M.; Ball, A.S. A quantitative PCR approach for quantification of functional gene involved in the degradation of polycyclic aromatic hydrocarbons in contaminated soils. *MethodsX* **2016**, *3*, 205–211. [CrossRef]
65. Gaby, J.C.; Buckley, D.H. A comprehensive evaluation of PCR primers to amplify the nifH gene of nitrogenase. *PLoS ONE* **2012**, *7*, e93883. [CrossRef]
66. Abe, J.; Popoola, O.; Popoola, A.; Ajenifuja, E.; Adebiyi, D. Application of Taguchi design method for optimization of spark plasma sintering process parameters for Ti-6Al-4V/h-BN binary composite. *Eng. Res. Express* **2019**, *1*, 025043. [CrossRef]
67. Dike, C.C.; Khudur, L.S.; Hakeem, I.G.; Rani, A.; Shahsavari, E.; Surapaneni, A.; Shah, K.; Ball, A.S. Biosolids-derived biochar enhances the bioremediation of diesel-contaminated soil. *J. Environ. Chem. Eng.* **2022**, *10*, 108633. [CrossRef]
68. Razmi, B.; Ghasemi-Fasaei, R.; Ronaghi, A.; Mostowfizadeh-Ghalamfarsa, R. Investigation of factors affecting phytoremediation of multi-elements polluted calcareous soil using Taguchi optimization. *Ecotoxicol. Environ. Saf.* **2021**, *207*, 111315. [CrossRef]
69. Zhang, Z.; Solaiman, Z.M.; Meney, K.; Murphy, D.V.; Rengel, Z. Biochars immobilize soil cadmium, but do not improve growth of emergent wetland species Juncus subsecundus in cadmium-contaminated soil. *J. Soils Sediments* **2013**, *13*, 140–151. [CrossRef]
70. Zheng, C.; Wang, X.; Liu, J.; Ji, X.; Huang, B. Biochar-assisted phytoextraction of arsenic in soil using *Pteris vittata* L. *Environ. Sci. Pollut. Res.* **2019**, *26*, 36688–36697. [CrossRef]
71. Jing, F.; Yang, Z.; Chen, X.; Liu, W.; Guo, B.; Lin, G.; Huang, R.; Liu, W. Potentially hazardous element accumulation in rice tissue and their availability in soil systems after biochar amendments. *J. Soils Sediments* **2019**, *19*, 2957–2970. [CrossRef]
72. Mohamed, I.; Zhang, G.-s.; Li, Z.-g.; Liu, Y.; Chen, F.; Dai, K. Ecological restoration of an acidic Cd contaminated soil using bamboo biochar application. *Ecol. Eng.* **2015**, *84*, 67–76. [CrossRef]
73. Barbafieri, M.; Dadea, C.; Tassi, E.; Bretzel, F.; Fanfani, L. Uptake of heavy metals by native species growing in a mining area in Sardinia, Italy: Discovering native flora for phytoremediation. *Int. J. Phytoremediation* **2011**, *13*, 985–997. [CrossRef] [PubMed]
74. Sun, W.; Shahrajabian, M.H.; Cheng, Q. Nitrogen fixation and diazotrophs–a review. *Rom. Biotechnol. Lett.* **2021**, *26*, 2834–2845. [CrossRef]
75. Ullah, A.; Mushtaq, H.; Ali, H.; Munis, M.F.H.; Javed, M.T.; Chaudhary, H.J. Diazotrophs-assisted phytoremediation of heavy metals: A novel approach. *Environ. Sci. Pollut. Res.* **2015**, *22*, 2505–2514. [CrossRef] [PubMed]
76. Shang, X.; Wu, S.; Liu, Y.; Zhang, K.; Guo, M.; Zhou, Y.; Zhu, J.; Li, X.; Miao, R. Rice husk and its derived biochar assist phytoremediation of heavy metals and PAHs co-contaminated soils but differently affect bacterial community. *J. Hazard. Mater.* **2024**, *466*, 133684. [CrossRef]
77. Huang, J.-H. Impact of microorganisms on arsenic biogeochemistry: A review. *Water Air Soil. Pollut.* **2014**, *225*, 1848. [CrossRef]
78. Xie, Z.; Luo, Y.; Wang, Y.; Xie, X.; Su, C. Arsenic resistance and bioaccumulation of an indigenous bacterium isolated from aquifer sediments of Datong Basin, Northern China. *Geomicrobiol. J.* **2013**, *30*, 549–556. [CrossRef]
79. Zhang, M.; Bai, S.H.; Tang, L.; Zhang, Y.; Teng, Y.; Xu, Z. Linking potential nitrification rates, nitrogen cycling genes and soil properties after remediating the agricultural soil contaminated with heavy metal and fungicide. *Chemosphere* **2017**, *184*, 892–899. [CrossRef]

MDPI AG
Grosspeteranlage 5
4052 Basel
Switzerland
Tel.: +41 61 683 77 34

Environments Editorial Office
E-mail: environments@mdpi.com
www.mdpi.com/journal/environments

www.ingramcontent.com/pod-product-compliance
Lightning Source LLC
Chambersburg PA
CBHW062034220926
51850CB00010B/2545
* 9 7 8 3 7 2 5 8 5 1 1 9 5 *